This book comes with access to more content online.
Quiz yourself and track your progress.

Register your book or ebook at
www.dummies.com/go/getaccess.

Select your product, and then follow the prompts
to validate your purchase.

You'll receive an email with your PIN and instructions.

Geometry Workbook

2nd Edition

by Mark Ryan

Geometry Workbook For Dummies®, 2nd Edition

Published by: **John Wiley & Sons, Inc.**, 111 River Street, Hoboken, NJ 07030-5774, www.wiley.com

For general information on our other products and services, please contact our Customer Care Department within the U.S. at 877-762-2974, outside the U.S. at 317-572-3993, or fax 317-572-4002. For technical support, please visit https://hub.wiley.com/community/support/dummies.

Wiley publishes in a variety of print and electronic formats and by print-on-demand. Some material included with standard print versions of this book may not be included in e-books or in print-on-demand. If this book refers to media that is not included in the version you purchased, you may download this material at http://booksupport.wiley.com. For more information about Wiley products, visit www.wiley.com.

Library of Congress Control Number: 2024946943

ISBN 978-1-394-27612-7 (pbk); ISBN 978-1-394-27616-5 (ebk); ISBN 978-1-394-27614-1 (ebk)

SKY10086579_100224

Table of Contents

Introduction

If you've already bought this book, then you have my undying respect and admiration (not to mention — cha ching — that with my royalty from the sale of this book, I can now afford, oh, say, half a cup of coffee). And if you're just thinking about buying it, well, what are you waiting for? Buying this book (and its excellent companion volume, *Geometry For Dummies*) can be an important first step on the road to gaining a solid grasp of a subject — and now I'm being serious — that is full of mathematical richness and beauty. By studying geometry, you take part in a long tradition going back at least as far as Pythagoras (one of the early, well-known mathematicians to study geometry, but certainly not the first). There is no mathematician, great or otherwise, who has not spent some time studying geometry.

I spend a great deal of time in this book explaining how to do geometry proofs. *Many* students have a lot of difficulty when they attempt their first proofs. I can think of a few reasons for this. First, geometry proofs, like the rest of geometry, have a spatial aspect that many students find challenging. Second, proofs lack the cut-and-dried nature of most of the math that students are accustomed to (in other words, with geometry proofs there are *way* more instances where there are many correct ways to proceed, and this takes some getting used to). And third, proofs are, in a sense, only half math. The other half is deductive logic — something new for most students, and something that has a significant *verbal* component. The good news is that if you practice the dozen or so strategies and tips for doing proofs presented in this book, you should have little difficulty getting the hang of it. These strategies and tips work like a charm and make many proofs much easier than they initially seem.

About This Book

Geometry Workbook For Dummies, like *Geometry For Dummies*, is intended for three groups of readers:

>> High school students (and possibly junior high students) taking a standard geometry course with the traditional emphasis on geometry proofs

>> The parents of geometry students

>> Anyone of any age who is curious about this interesting subject, which has fascinated both mathematicians and laypeople for well over two thousand years

Whenever possible, I explain geometry concepts and problem solutions with a minimum of technical jargon. I take a common-sense, street-smart approach when explaining mathematics, and I try to avoid the often stiff and formal style used in too many textbooks. You get answer explanations for every practice problem. And with proofs, in addition to giving you the steps of the solutions, I show you the thought process behind the solutions. I supplement the problem explanations with tips, shortcuts, and mnemonic devices. Often, a simple tip or memory trick can make learning and retaining a new, difficult concept much easier. The pages here should contain enough blank space to allow you to write out your solutions right in the book.

Conventions Used in This Book

This book uses certain conventions:

>> Variables are in *italics*.

>> Important math terms are often in *italics* and are defined when necessary. These terms may be **bolded** when they appear as keywords within a bulleted list. Italics are also used for emphasis.

>> As in most geometry books, figures are not necessarily drawn to scale.

>> Extra-hard problems are marked with an asterisk. Don't try these problems on an empty stomach!

For all proof problems, don't assume that the number of blank lines (where you'll put your solutions) corresponds exactly to the number of steps needed for the proof.

How to Use This Book

Like all *For Dummies* books, you can use this book as a reference. You don't need to read it cover to cover or work through all problems in order. You may need more practice in some areas than others, so you may choose to do only half of the practice problems in some sections, or none at all.

However, as you'd expect, the order of the topics in *Geometry Workbook For Dummies* roughly follows the order of a traditional high school geometry course. You can, therefore, go through the book in order, using it to supplement your coursework. If I do say so myself, I expect you'll find that many of the explanations, methods, strategies, and tips in this book will make problems you found difficult or confusing in class seem much easier.

I give hints for many problems, but if you want to challenge yourself, you may want to cover them up and attempt the problem without the hint.

And if you get stuck while doing a proof, you can try reading a little bit of the "game plan" or the solution to the proof. These aids are in the solutions section at the end of every chapter. But don't read too much at first. Read a small amount and see whether it gives you any ideas. Then, if you're still having trouble, read a little more.

Foolish Assumptions

As William Shakespeare said, "A fool thinks himself to be wise, but a wise man knows himself to be a fool." Here's what I'm assuming about you — fool that I am.

>> You're no slouch — and therefore, you have at least some faint glimmer of curiosity about geometry (or maybe you're totally, stark raving mad with desire to learn the subject?). How could people possibly have no curiosity at all about geometry, assuming they're not in a coma? You are literally surrounded by shapes, and every shape involves geometry.

>> You haven't forgotten basic algebra. You need very little algebra for geometry, but you do need some. Even if your algebra is a bit rusty, I doubt you'll have any trouble with the algebra in this book: solving simple equations, using simple formulas, doing square roots, and so on.

>> You're willing to invest some time and effort in doing these practice problems. With geometry — as with anything — practice makes perfect, and practice sometimes involves struggle. But that's a good thing. Ideally, you should give these problems your best shot before you turn to the solutions. Reading through the solutions can be a good way to learn, but you'll usually remember more if you first push yourself to solve the problems on your own — even if that means going down a few dead ends.

Icons Used in This Book

Look for the following icons to quickly spot important information:

Next to this icon are definitions of geometry terms, explanations of geometry principles, and a few things you should know from algebra. You often use geometry definitions in the reason column of two-column proofs.

This icon is next to all example problems — duh.

This icon gives you shortcuts, memory devices, strategies, and so on.

Ignore these icons, and you may end up doing lots of extra work and maybe getting the wrong answer — and then you could fail geometry, become unpopular, and lose any hope of becoming homecoming queen or king. Better safe than sorry, right?

This icon identifies the theorems and postulates that you'll use to form the chain of logic in geometry proofs. You use them in the reason column of two-column proofs. A *theorem* is an if-then statement, like "if angles are supplementary to the same angle, then they are congruent." You use *postulates* basically the same way that you use theorems. The difference between them is sort of a mathematical technicality (which I wouldn't sweat if I were you).

Beyond the Book

You have online access to hundreds of geometry practice problems to supplement what's covered in the book. To gain access to this online practice material, all you have to do is register. Just follow these simple steps:

1. **Register your book or e-book at** Dummies.com **to get your personal identification number (PIN).**

   Go to www.dummies.com/go/getaccess.

2. **Choose your product from the drop-down list on that page.**

3. **Follow the prompts to validate your product.**

4. **Check your email for a confirmation message that includes your PIN and instructions for logging in.**

   If you don't receive this email within two hours, please check your spam folder before contacting us through our support website at http://support.wiley.com or by phone at +1 (877) 762-2974.

Where to Go from Here

You can go

» To Chapter 1

» To whatever chapter contains the concepts you need to practice

» To *Geometry For Dummies* for more in-depth explanations

» To the movies

» To the beach

» Into your geometry final to kick some @#%$!

» Then on to bigger and better things

1

Getting Started with Geometry

Chapter 1

Introducing Geometry and Geometry Proofs

I n this chapter, you get started with some basics about geometry and shapes, a couple points about deductive logic, and a few introductory comments about the structure of geometry proofs. Time to get started!

What Is Geometry?

What is geometry?! C'mon, everyone knows what geometry is, right? *Geometry* is the study of shapes: circles, triangles, rectangles, pyramids, and so on. Shapes are all around you. The desk or table where you're reading this book has a shape. You can probably see a window from where you are, and it's probably a rectangle. The pages of this book are also rectangles. Your pen or pencil is roughly a cylinder (or maybe a right hexagonal prism — see Part 5 for more on solid figures). Your shirt may have circular buttons. The bricks of a brick house are right rectangular prisms. Shapes are ubiquitous — in our world, anyway.

For the philosophically inclined, here's an exercise that goes *way* beyond the scope of this book: Try to imagine a world — some sort of different universe — where there aren't various objects with different shapes. (If you're into this sort of thing, check out *Philosophy For Dummies*.)

Making the Right Assumptions

Okay, so geometry is the study of shapes. And how can you tell one shape from another? From the way it looks, of course. But — this may seem a bit bizarre — when you're studying geometry, you're sort of *not* supposed to rely on the way shapes look. The point of this strange treatment of geometric figures is to prohibit you from claiming that something is true about a figure merely because it looks true, and to force you, instead, to *prove* that it's true by airtight, mathematical logic.

When you're working with shapes in any other area of math, or in science, or in, say, architecture or design, paying attention to the way shapes look is very important: their proportions, their angles, their orientation, how steep their sides are, and so on. Only in a geometry course are you supposed to ignore to some degree the appearance of the shapes you study. (I say "to some degree" because, in reality, even in a geometry course — or when using this book — it's still quite useful most of the time to pay attention to the appearance of shapes.)

WARNING

When you look at a diagram in this or any geometry book, **you *cannot* assume any of the following just from the appearance of the figure.**

>> **Right angles:** Just because an angle looks like an exact 90° angle, that doesn't necessarily mean it is one.

>> **Congruent angles:** Just because two angles look the same size, that doesn't mean they really are. (As you probably know, *congruent* [symbolized by ≅] is a fancy word for "equal" or "same size.")

>> **Congruent segments:** Just like with angles, you can't assume segments are the same length just because they appear to be.

>> **Relative sizes of segments and angles:** Just because, say, one segment is drawn to look longer than another in some diagram, it doesn't follow that the segment really is longer.

Sometimes size relationships are marked on the diagram. For instance, a small L-shaped mark in a corner means that you have a right angle. Tick marks can indicate congruent parts. Basically, if the tick marks match, you know the segments or angles are the same size.

You can assume pretty much anything not on this list; for example, if a line looks straight, it really is straight.

Before doing the following problems, you may want to peek ahead to Chapters 4 and 6 if you've forgotten or don't know the names of various triangles and quadrilaterals.

Q. What can you assume and what can't you assume about *SIMON*?

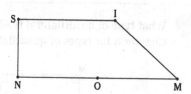

A. You *can* assume that

- $\overline{MN}$ (line segment *MN*) is straight; in other words, there's no bend at point *O*.

  Another way of saying the same thing is that ∠*MON* is a *straight angle* or a 180° angle.

- $\overline{NS}$, $\overline{SI}$, and $\overline{IM}$ are also straight as opposed to curvy.

- Therefore, *SIMON* is a quadrilateral because it has four straight sides.

  (If you couldn't assume that $\overline{MN}$ is straight, there could actually be a bend at point *O* and then *SIMON* would be a pentagon, but that's not possible.)

  That's about it for what you can assume. If this figure were anywhere else other than a geometry book, you could safely assume all sorts of other things — including that *SIMON* is a trapezoid. But this *is* a geometry book, so you *can't* assume that. You also *can't* assume that

- ∠*S* and ∠*N* are right angles.

- ∠*I* is an obtuse angle (an angle greater than 90°).

- ∠*M* is an acute angle (an angle less than 90°).

- ∠*I* is greater than ∠*M* or ∠*S* or ∠*N*, and ditto for the relative sizes of other angles.

- $\overline{NS}$ is shorter than $\overline{SI}$ or $\overline{MN}$, and ditto for the relative lengths of the other segments.

- *O* is the midpoint of $\overline{MN}$.

- $\overline{SI}$ is parallel to $\overline{MN}$.

The "real" *SIMON* — weird as it seems — could actually look like this:

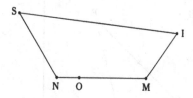

What type of quadrilateral is *AMER? Note:* See Chapter 6 for types of quadrilaterals.

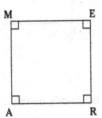 What type of quadrilateral is *IDOL?*

 Use the figure to answer the following questions (Chapter 4 can fill you in on triangles):

a. Can you assume that the triangles are congruent?

b. Can you conclude that △*ABC* is acute? Obtuse? Right? Isosceles (with at least two equal sides)? Equilateral (with three equal sides)?

c. Can you conclude that △*DEF* is acute? Obtuse? Right? Isosceles? Equilateral?

d. What can you conclude about the length of $\overline{EF}$?

e. Might ∠*D* be a right angle?

f. Might ∠*F* be a right angle?

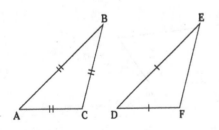

 Can you assume or conclude

a. △*ABC* ≅ △*WXY*?

b. △*ABD* ≅ △*CBD*?

c. △*ABD* ≅ △*WXZ*?

d. △*ABC* is isosceles?

e. *D* is the midpoint of $\overline{AC}$?

f. *Z* is the midpoint of $\overline{WY}$?

g. $\overline{BD}$ is an altitude (height) of △*ABC*?

h. ∠*ADB* is supplementary to ∠*CDB* (that is: ∠*ADB* + ∠*CDB* = 180°)?

i. △*XYZ* is a right triangle?

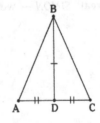

If-Then Logic: If You Bought This Book, Then You Must Love Geometry!

Geometry *theorems* (and their cousins, *postulates*) are basically statements of geometrical truth, like "All radii of a circle are congruent." As you can see in this section and in the rest of the book, theorems (and postulates) are the building blocks of proofs. (I may get hauled over by the geometry police for saying this, but if I were you, I'd just glom theorems and postulates together into a single group because, for the purposes of doing proofs, they work the same way. Whenever I refer to theorems, you can safely read it as "theorems and postulates.")

Geometry theorems can all be expressed in the form, "*If* blah blah blah, *then* blah blah blah," like "If two angles are right angles, then they are congruent" (although theorems are often written in some shorter way, like "All right angles are congruent"). You may want to flip through the book looking for theorem icons to get a feel for what theorems look like.

WARNING

An important thing to note here is that **the reverse of a theorem is not necessarily true**. For example, the statement, "If two angles are congruent, then they are right angles," is false. When a theorem does work in both directions, you get two separate theorems, one the reverse of the other.

The fact that theorems are not generally reversible should come as no surprise. Many ordinary statements in *if-then* form are, like theorems, not reversible: "If something's a ship, then it's a boat" is true, but "If something's a boat, then it's a ship" is false, right? (It might be a canoe.)

Geometry definitions (like all definitions), however, are reversible. Consider the definition of *perpendicular:* perpendicular lines are lines that intersect at right angles. Both if-then statements are true: 1) "If lines are perpendicular, then they intersect at right angles," and 2) "If lines intersect at right angles, then they are perpendicular." When doing proofs, you'll have the occasion to use both forms of many definitions.

EXAMPLE

Q. Read through some theorems.

   a. Give an example of a theorem that's not reversible and explain why the reverse is false.

   b. Give an example of a theorem whose reverse is another true theorem.

A. A number of responses work, but here's how you could answer:

   a. "If two angles are vertical angles, then they are congruent." The reverse of this theorem is obviously false. Just because two angles are the same size, it does not follow that they must be vertical angles. (When two lines intersect and form an X, vertical angles are the angles straight across from each other — turn to Chapter 2 for more info.)

   b. Two of the most important geometry theorems are a reversible pair: "If two sides of a triangle are congruent, then the angles opposite those sides are congruent" and "If two angles of a triangle are congruent, then the sides opposite those angles are congruent." (For more on these isosceles triangle theorems, check out Chapter 5.)

5 Give two examples of theorems that are not reversible and explain why the reverse of each is false. *Hint:* Flip through this book or your geometry textbook and look at various theorems. Try reversing them and ask yourself whether they still work.

6 Give two examples of theorems that work in both directions. *Hint:* See the hint for question 5.

What's a Geometry Proof?

Many students find two-column geometry proofs difficult, but they're really no big deal once you get the hang of them. Basically, they're just arguments like the following, in which you brilliantly establish that your Labradoodle, Fifi, will not lay any eggs on the Fourth of July:

1. Fifi is a Labradoodle.

2. Therefore, Fifi is a dog, because all Labradoodles are dogs.

3. Therefore, Fifi is a mammal, because all dogs are mammals.

4. Therefore, Fifi will never lay any eggs, because mammals don't lay eggs (okay, okay . . . except for platypuses and spiny anteaters, for you monotreme-loving nitpickers out there).

5. Therefore, Fifi will not lay any eggs on the Fourth of July, because if she will never lay any eggs, she can't lay eggs on the Fourth of July.

In a nutshell: Labradoodle → dog → mammal → no eggs → no eggs on July 4. It's sort of a domino effect. Each statement knocks over the next till you get to your final conclusion.

EXAMPLE

Check out Figure 1-1 to see what this argument or proof looks like in the standard two-column geometry proof format.

Given: Fifi is a Labradoodle.
Prove: Fifi will not lay eggs on the Fourth of July.

Statements (or Conclusions) These are the specific claims you make.	Reasons (or Justifications) These are the general rules you use to justify your claims. If after each claim you made, I said, "How do you know?" your response to me goes in this column.
I claim...	**How do I know?**
1) Fifi is a Labradoodle.	1) Because it was given as a fact.
2) Fifi is a dog.	2) Because all Labradoodles are dogs.
3) Fifi is a mammal.	3) Because all dogs are mammals.
4) Fifi doesn't lay eggs.	4) Because mammals don't lay eggs.
5) Fifi will not lay eggs on the Fourth of July.	5) Because something that doesn't lay eggs can't lay eggs on the Fourth of July.

FIGURE 1-1: A standard two-column proof listing statements and reasons.

TIP

Note that the left-hand column contains *specific* facts (about one particular dog, Fifi), **while the right-hand column contains** *general* **principles** (about dogs in general or mammals in general). This format is true of all geometry proofs.

Now look at the very same proof in Figure 1-2; this time, the reasons appear in *if-then* form. When reasons are written this way, you can see how the chain of logic flows.

REMEMBER

In a two-column proof, the idea or ideas in the *if* part of each reason must come from the statement column somewhere *above* the reason; and the single idea in the *then* part of the reason must match the idea in the statement on *the same line* as the reason. This incredibly important flow-of-logic structure is shown with arrows in the following proof.

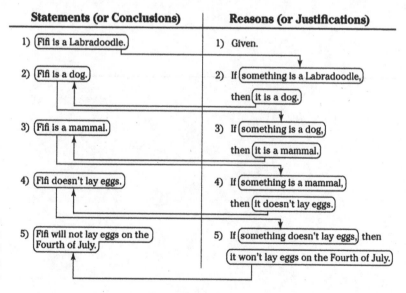

FIGURE 1-2: A proof with the reasons written in if-then form.

In the preceding proof, each *if* clause uses only a single idea from the statement column. However, as you can see in the following practice problem, you often have to use more than one idea from the statement column in an *if* clause.

 In the following facetious and somewhat fishy proof, fill in the missing reasons in *if-then* form and show the flow of logic as I illustrate in Figure 1-2.

Given: You forgot to set your alarm last night.

          You've already been late for school twice this term.

Prove: You will get a detention at school today.

Note: To complete this "proof," you need to know the school's late policy: A student who is late for school three times in one term will be given a detention.

Statements (or Conclusions)	Reasons (or Justifications)
1) I forgot to set my alarm last night.	1) Given.
2) I will wake up late.	2)
3) I will miss the bus.	3)
4) I will be late for school.	4)
5) I've already been late for school twice this term.	5) Given.
6) This will be the third time this term I'll have been late.	6)
7) I'll get a detention at school today.	7)

Solutions

1. *AMER* looks like a square, but you can't conclude that because you can't assume the sides are equal. You do know, however, that the figure is a rectangle because it has four sides and four right angles.

2. *IDOL* also looks like a square, and again, like with question 1, you can't conclude that, but this time you can't conclude that because you can't assume that the angles are right angles. But because you do know that *IDOL* has four equal sides, you know that it's a rhombus.

3. Here are the answers (flip to Chapter 4 if you need to go over triangle classification):

   a. No. The triangles look congruent, but you're not allowed to assume that.

   b. The tick marks tell you that $\triangle ABC$ is equilateral. It is, therefore, an acute triangle and an isosceles triangle. It is neither a right triangle nor an obtuse triangle.

   c. The tick marks tell you that $\triangle DEF$ is isosceles and that, therefore, it is not scalene. That's all you can conclude. It may or may not be any of the other types of triangles.

   d. Nothing. $\overline{EF}$ could be the longest side of the triangle, or the shortest, or equal to the other two sides. And it may or may not have the same length as $\overline{BC}$.

   e. Yes. $\angle D$ might be a right angle, though you can't assume that it is.

   f. No. (If you got this question right, give yourself a pat on the back.) If $\angle F$ were a right angle, $\triangle DEF$ would be a right triangle with $\overline{DE}$ its hypotenuse. But $\overline{DE}$ is the same length as $\overline{DF}$, and the hypotenuse of a right triangle has to be the triangle's longest side.

4. Here are the answers:

   a. No. The triangles might not be congruent in any number of ways. For example, you know nothing about the length of $\overline{ZY}$, and if $\overline{ZY}$ were, say, a mile long, the triangles would obviously not be congruent.

   b. No. The triangles would be congruent only if $\angle ADB$ and $\angle CDB$ were right angles, but you don't know that. Point B is free to move left or right, changing the measures of $\angle ADB$ and $\angle CDB$.

   c. No. You don't know that $\angle ADB$ is a right angle.

   d. No. The figure *looks* isosceles, but you're not allowed to assume that $\overline{AB} \cong \overline{CB}$.

   e. Yes. The tick marks show it.

   f. No. Like with part *a*, you know nothing about the length of $\overline{ZY}$.

   g. No. You can't assume that $\overline{BD} \perp \overline{AC}$ (the upside-down *T* means "is perpendicular to").

   h. Yes. You *can* assume that $\overline{AC}$ is straight and that $\angle ADC$ is 180°; therefore, $\angle ADB$ and $\angle CDB$ must add up to 180°.

   i. Yes. $\angle WZY$ is 180° and $\angle WZX$ is 90°, so $\angle YZX$ must also be 90°.

5. Answers vary. One example is "If angles are complementary to the same angle, then they're congruent." The reverse of this is false because many angles, like obtuse angles, do not have complements (obtuse angles are already bigger than 90°, so you can't add another angle to them to get a right angle).

(6) Answers vary. Any of the parallel line theorems in Chapter 2 makes a good answer. For example, "If two parallel lines are cut by a transversal, then alternate interior angles are congruent." Both this theorem and its reverse are true. To wit (in abbreviated form): "If lines are parallel, then alternate interior angles are congruent," and "If alternate interior angles are congruent, then lines are parallel."

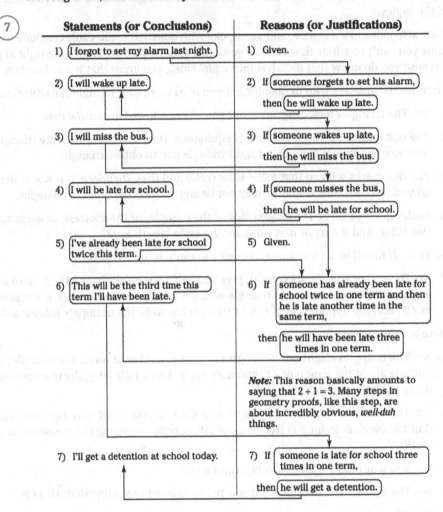

(7)

Statements (or Conclusions)	Reasons (or Justifications)
1) I forgot to set my alarm last night.	1) Given.
2) I will wake up late.	2) If someone forgets to set his alarm, then he will wake up late.
3) I will miss the bus.	3) If someone wakes up late, then he will miss the bus.
4) I will be late for school.	4) If someone misses the bus, then he will be late for school.
5) I've already been late for school twice this term.	5) Given.
6) This will be the third time this term I'll have been late.	6) If someone has already been late for school twice in one term and then he is late another time in the same term, then he will have been late three times in one term.
	Note: This reason basically amounts to saying that 2 + 1 = 3. Many steps in geometry proofs, like this step, are about incredibly obvious, *well-duh* things.
7) I'll get a detention at school today.	7) If someone is late for school three times in one term, then he will get a detention.

I hope it goes without saying that this is *not* an airtight, mathematical proof.

Chapter **2**

Points, Segments, Lines, Rays, and Angles

I n this chapter, you first review the building blocks of geometry: points, segments, lines, rays, and angles. Then I go over some terms related to those objects: midpoint, bisection, and trisection; parallel and perpendicular lines; right, acute, and obtuse angles; complementary and supplementary angles; and vertical angles. You'll get the hang of these things working through the practice problems.

Hammering Out Basic Definitions

REMEMBER

You probably already know what the following things are, but here are their definitions and undefinitions anyway. That's right — I said *undefinitions*. Technically, *point* and *line* are undefined terms, so the first two "definitions" that follow aren't technically definitions. But if I were you, I wouldn't sweat this technicality.

» **Point:** You know, like a dot except that it actually has no size at all. Or, you could say that it's infinitely small. (That's pretty small, eh? But even *"infinitely* small" makes a point sound larger than it really is.)

» **Line:** A line's like a thin, straight wire. (Actually, it's infinitely thin or, even better, it has *no width at all* — nada.) Don't forget that it goes on forever in both directions, which is why you use the little double-headed arrow as in $\overleftrightarrow{AB}$ (read as "line *AB*"; this is the line that goes through points *A* and *B*). Because lines go on forever, no matter how you tilt them or how good your shoehorn is, you can't fit them in the universe.

>> **Line segment** or just **segment:** A segment is a section of a line that has two endpoints. If it goes from *C* to *D*, you call it "segment *CD*" and write it like $\overline{CD}$. (You can also call it and write it $\overline{DC}$.)

Note: CD without the segment bar over it indicates the *length* of the segment as opposed to the segment itself.

>> **Ray:** A ray is a section of a line (sort of half a line) that has one endpoint and goes on forever in the other direction. If its endpoint is point *M* and the ray goes through point *N* and then past it forever, you call the half-line "ray *MN*" and write $\overrightarrow{MN}$. The endpoint always comes first.

>> **Angle:** Two rays with the same endpoint form an angle. The common endpoint is called the *vertex* of the angle. An *acute angle* is less than 90°; a *right angle* is, of course, a 90° angle; an *obtuse angle* has a measure greater than 90°; and a *straight angle* has a measure of 180° (which is kinda weird, because a 180° angle looks just like a line or a segment like ∠*ACE* in the example in the next section).

Note: Technically, angles go on forever, and their sides are rays that go on forever. This is the case even when an angle in a figure has segments for its sides instead of rays. (It's like the rays are really there even though they're not drawn.)

Looking at Union and Intersection Problems

REMEMBER

And now for something completely different. **The *intersection* (∩) of two geometric objects is where they overlap or touch. The *union* (∪) of two objects contains all of each object including the overlapping portion (if any).**

EXAMPLE

Q. For the figure on the right, determine the following and write your answer in as many ways as possible.

a. $\overrightarrow{AE} \cap \overrightarrow{CA}$

b. $\overrightarrow{AE} \cup \overrightarrow{CA}$

c. ∠*BDE* ∩ $\overrightarrow{ED}$

d. ∠*BDE* ∪ $\overrightarrow{DE}$

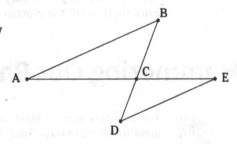

A. Here's how this problem shapes up:

a. $\overrightarrow{AE} \cap \overrightarrow{CA} = \overline{AC}$ or $\overline{CA}$

b. $\overrightarrow{AE} \cup \overrightarrow{CA} = \overline{AC}$ or $\overline{CA}$ or $\overrightarrow{AE}$ or $\overrightarrow{EA}$ or $\overrightarrow{CE}$ or $\overrightarrow{EC}$

Tip: If you find some of these union and intersection problems tricky, you're not alone. Here's a great way to do them or to think about them. Imagine that the first object is colored blue and the second, yellow (or you can actually color them). Blue and yellow make green, right? So, wherever you see (or imagine) green, that's the intersection. And the union contains anything that's blue or yellow or green. Another way to do these problems is to trace over each object. Wherever you traced twice, that's the intersection. And wherever you did any tracing (once or twice), that's the union.

Remember: However you do these problems, lines, rays, and angles go on forever even if the diagram makes it look like they end.

c. $\angle BDE \cap \overleftrightarrow{ED} = \overrightarrow{DE}$

    If $\angle BDE$ is blue and $\overleftrightarrow{ED}$ is yellow, then $\overrightarrow{DE}$ will be green.

d. $\angle BDE \cup \overrightarrow{DE} = \angle BDE$ or $\angle EDB$ or $\angle CDE$ or $\angle EDC$

Note that sometimes the answer to a union or intersection problem is one of the original objects.

Use the following figure to answer problems 1 to 6.

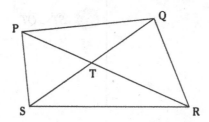

1. $\overline{ST} \cap \overline{TQ}$

2. $\overline{ST} \cup \overline{PT}$

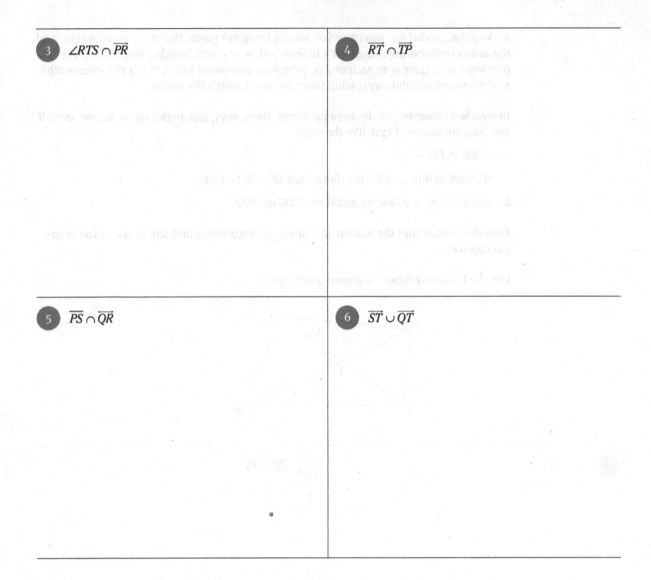

3 $\angle RTS \cap \overline{PR}$

4 $\overline{RT} \cap \overline{TP}$

5 $\overline{PS} \cap \overline{QR}$

6 $\overline{ST} \cup \overline{QT}$

Uncovering More Definitions

In the sections that follow, I give you roughly ten more definitions of important geometry terms. You'll get practice using these terms in this chapter's problems, and then you'll use these terms throughout the rest of the book.

Division in the Ranks: Bisection and Trisection

REMEMBER

In this section, you practice something you've understood almost since you first rode a bicycle or tricycle: cutting things in half or in thirds. This geometry is kids' stuff. Check out the following definitions.

- ▸▸ **Segment bisection** and **midpoint:** A point, segment, ray, or line that divides a segment into two congruent segments *bisects* the segment. The point of bisection is called the *midpoint* of the segment. The midpoint, obviously, cuts the segment in half.

- ▸▸ **Segment trisection:** Two things (points, segments, rays, lines, or any combination of these) that divide a segment into three congruent segments *trisect* the segment. The points of trisection are called — get this — the *trisection points* of the segment.

- ▸▸ **Angle bisection:** A ray that cuts an angle into two congruent angles *bisects* the angle. It's called the *bisector* of the angle, or the *angle bisector*.

- ▸▸ **Angle trisection:** Two rays that divide an angle into three congruent angles *trisect* the angle. They're called *trisectors* of the angle, or *angle trisectors*.

EXAMPLE

Q. For the triangle on the right, given that $\overrightarrow{CD}$ bisects $\angle ACB$:

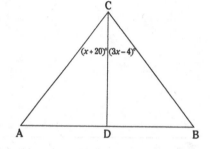

  a. Find the measure of $\angle BCD$.

  b. Other than the fact that $\angle ACD \cong \angle BCD$, can you conclude anything else about this figure?

A. Given that $\overrightarrow{CD}$ bisects $\angle ACB$:

  a. You can find the measure of $\angle BCD$ in two steps. First, because $\overrightarrow{CD}$ bisects $\angle ACB$, $\angle ACD \cong \angle BCD$, so you can set them equal to each other and solve for x:

$$x + 20 = 3x - 4$$
$$-2x = -24$$
$$= 12$$

Now plug 12 into the measure of $\angle BCD$ to get your answer:

$$\angle BCD = 3x - 4$$
$$= 3 \cdot 12 - 4$$
$$= 32°$$

  b. Other than the fact that $\angle ACD \cong \angle BCD$, you can conclude nothing else.

Don't jump to conclusions based on the appearance of figures. For this problem, you know only that $\overrightarrow{CD}$ bisects an *angle* ($\angle ACB$). You *cannot* conclude that $\overrightarrow{CD}$ bisects the base of the triangle, and therefore you don't know whether D is the midpoint of $\overline{AB}$. You also can't conclude that $\triangle ABC$ has been cut in half. And you can't say that $\overline{AC} \cong \overline{BC}$ or that $\angle A \cong \angle B$. Finally, you can't conclude that $\angle ADC$ and $\angle BDC$ are right angles.

7 On this number line, Q and R trisect $\overline{PS}$. What are the coordinates of Q and R?

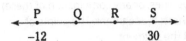

P Q R S

−12   30

 8 Given that $\angle 1 = 4x$, $\angle 2 = x + 9$, and $\angle 3 = 5x - 7$, is $\angle STU$ trisected?

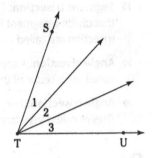

9 $\overrightarrow{NP}$ and $\overrightarrow{NQ}$ divide right $\angle MNO$ into $\angle MNP$, $\angle PNQ$, and $\angle QNO$, whose measures are in the ratio $4:5:6$. Determine the measure of $\angle PNO$.

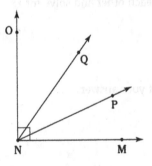

*10 Given: $\overline{BD}$ and $\overline{BE}$ trisect $\overline{AC}$; $\overline{AD}$ and $\overline{DE}$ have lengths as shown.

a. Determine DC (the length of the segment).

b. Can you conclude that $\angle 1 \cong \angle 2$? That $\angle 1 \cong \angle 3$?

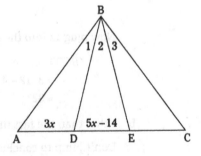

Perfect Hilarity for Perpendicularity

You're surrounded by perpendicular things: floors are perpendicular to walls, sides of rectangular shapes are perpendicular, the majority of streets that cross are perpendicular, and so on. In this section, you practice problems involving perpendicular lines (and rays and segments). I'm also going to give you the definition of *parallel*. Like perpendicular things, you also see parallel things every day: a ceiling is parallel to the floor, the top and bottom edges of this book are parallel, two lines of words on this page are parallel, and so on. I thought I'd give you this definition now mainly because perpendicular and parallel make a nice pair of geometry terms, but you won't use parallel lines till Chapter 6.

REMEMBER

Lines, rays, or segments that form a right angle are *perpendicular.* The symbol for perpendicularity is ⊥. (Note that you say that lines, rays, or segments are perpendicular and that an angle is a right angle; you do *not* say that an angle is perpendicular.)

REMEMBER

Lines, rays, or segments that run along in the same direction and never cross — like two railroad tracks — are *parallel.* The symbol for parallel is ∥. If lines $\overleftrightarrow{AB}$ and $\overleftrightarrow{CD}$ are parallel, you'd write $\overleftrightarrow{AB} \parallel \overleftrightarrow{CD}$.

EXAMPLE

Q. In the figure on the right, $\overline{BA} \perp \overline{BC}$, $\angle 1 \cong \angle 3$, and $\angle 2$ is three times as big as $\angle 1$. Find the measure of $\angle 2$.

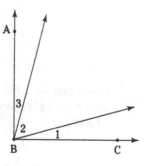

A. Because the rays are perpendicular, $\angle ABC$ is a right angle and thus measures 90°. $\angle 1$ and $\angle 3$ are equal, so you can set them both equal to x. $\angle 2$ is three times as big as $\angle 1$, so its measure is $3x$. Now you have three angles, $\angle 1$, $\angle 2$, and $\angle 3$, whose measures (x, $3x$, and x) must add up to 90. Thus,

$$x + 3x + x = 90$$
$$5x = 90$$
$$x = 18$$

Now, plugging 18 into $3x$ gives you the measure of $\angle 2$:

$$3(18) = 54°$$

11 In the following figure:

   a. Find ∠BFC given that ∠DFE measures 25°, that $\overleftrightarrow{AE} \perp \overleftrightarrow{FC}$, and that $\overleftrightarrow{FB} \perp \overleftrightarrow{FD}$.

   b. What two objects form the sides of ∠BFC?

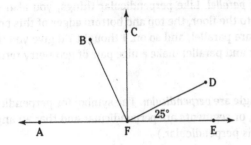

12 Given that $\overleftrightarrow{AD} \perp \overleftrightarrow{BE}$, ∠DGC measures 10°, and ∠BGC is four times as large as ∠AGF, find the measure of ∠FGE.

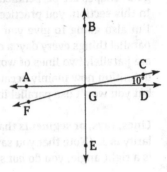

13 Given that $\overleftrightarrow{AF} \perp \overleftrightarrow{EH}$, that $\overrightarrow{BG}$ bisects ∠FIH, and that $\overrightarrow{IC}$ and $\overrightarrow{ID}$ trisect ∠BIE, find the measure of ∠BID.

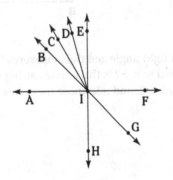

14 In the following figure, $\overline{RG} \perp \overline{RY}$, $\overline{RG} \perp \overline{GA}$, and $\overline{RY} \perp \overline{LN}$.

   a. Name the angles you know are right angles.

   b. Can you conclude that ∠ANL is a right angle?

   c. What's ∠GRY ∩ $\overline{YL}$?

   d. What's ∠GRY ∩ $\overline{LY}$?

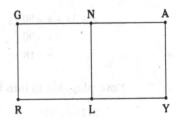

You Complete Me: Complementary and Supplementary Angles

Here are the definitions of two terms I have a feeling you've seen before. For those of you who (like me) like mnemonic devices, here's one for these terms. It's a bit lame, but better than nothing: In the following definitions, note that the terms are in alphabetical order and the numbers are in numerical order ("right" and "straight" are also in alphabetical order).

REMEMBER

» **Complementary angles:** Two angles whose sum is 90° (or a right angle)

» **Supplementary angles:** Two angles whose sum is 180° (or a straight angle)

EXAMPLE

Q. $\angle BQC$ is complementary to $\angle CQD$
$\angle BQC$ is supplementary to $\angle AQE$
$m\angle CQD + m\angle AQE = 200°$

Find: $m\angle BQC$

A. Set $m\angle BQC$ equal to x. Then, because $\angle BQC$ and $\angle CQD$ are complementary, $m\angle CQD = 90 - x$, and, because $\angle BQC$ and $\angle AQE$ are supplementary, $m\angle AQE = 180 - x$. These angles add to 200°. Thus,

$$(90 - x) + (180 - x) = 200$$
$$-2x = -70$$
$$x = 35$$

That's it. $m\angle BQC = 35°$

---

15 Given: $\angle 1 = 25°$

$\angle 2 = 90°$

$\angle 4$ is complementary to $\angle 6$

Find: The measures of angles 3 through 9

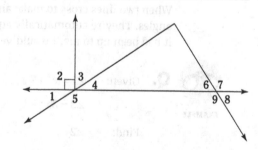

16 The supplement of an angle is 20° greater than twice the angle's complement. Find the angle's measure.

X Marks the Spot: Vertical Angles

Don't ask me how they came up with the term *vertical angles*, because these angles have nothing to do with the ordinary meaning of *vertical* (you know, as in vertical and horizontal). Go figure. When two lines cross to make an X, the two angles on opposite sides of the X are called vertical angles. They're automatically equal. As you can see, every X has two pairs of vertical angles. If it had been up to me, I would've called them *x-angles* or *cross angles*.

EXAMPLE

Q. Given: $\angle 1 = x^2 + 7$

$\angle 3 = 3x^2 - 1$

Find: $\angle 2$

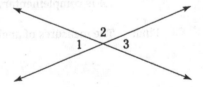

A. $\angle 1$ and $\angle 3$ are vertical angles and are thus equal, so set them equal to each other and solve for x:

$$3x^2 - 1 = x^2 + 7$$
$$2x^2 = 8$$
$$x^2 = 4$$
$$x = \pm 2$$

Plug $x = 2$ into the measure of $\angle 1$:

$$\angle 1 = x^2 + 7$$
$$= 2^2 + 7$$
$$= 11$$

Figure the measure of $\angle 2$:

$$\angle 2 = 180 - \angle 1$$
$$= 180 - 11$$
$$= 169$$

Ordinarily, you'd now want to plug $x = -2$ into the measure of $\angle 1$, repeat the last two steps, and maybe get a second answer for $\angle 2$. But in this particular problem, you don't have to do that, because regardless of whether $x = 2$ or -2, everything comes out the same (squaring a negative gives you a positive). In general, however, you have to plug in each solution for x separately.

Remember: Segments and angles must, of course, have *positive* lengths or measures. So, if you plug an x-value into a segment or angle and your answer is zero or negative, reject that x-value.

Warning: Be careful, however, not to reject x-values simply because they are zero or negative. The segments and angles, not x, must be positive. There are plenty of problems in which a *negative* solution for x gives you a *positive* answer for a segment or angle, and vice versa.

 Use the figures to answer the following questions.

a. Is this possible?

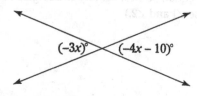

b. Is this possible?

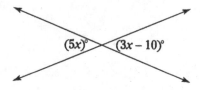

18 Solve for $\angle AQB$ and $\angle DQC$.

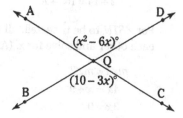

Solutions

(1) $\overline{ST} \cap \overline{TQ}$ = point T

(2) $\overline{ST} \cup \overline{PT}$ = ?

Did you have ∠STP for this one? Nope! That incorrect answer would mean an angle made up of two rays that go on forever. But this problem involves segments, not rays. Technically, this union is not an angle. And there is no nice, simple name for this thing. You can't really write it any more simply than you see it in the original problem: $\overline{ST} \cup \overline{PT}$.

(3) $\angle RTS \cap \overrightarrow{PR} = \overrightarrow{TR}$

If you trace over ∠RTS (remembering that it goes out past R forever) and then over $\overrightarrow{PR}$, you trace twice over $\overrightarrow{TR}$, the ray that begins at T and goes out forever past R.

(4) $\overrightarrow{RT} \cap \overrightarrow{TP} = \overline{TP}$

(5) $\overrightarrow{PS} \cap \overrightarrow{QR} = \varnothing$ (the empty set)

They don't overlap anywhere.

(6) $\overrightarrow{ST} \cup \overrightarrow{QT} = \overline{SQ}$ or $\overline{QS}$ or $\overrightarrow{ST}$ or $\overrightarrow{TS}$ or $\overrightarrow{TQ}$ or $\overrightarrow{QT}$

(7) You can solve this problem in two steps:

$\overline{PS}$ is trisected, so determine PS and then divide that by 3:

$$PS = 30 - (-12) = 42$$
$$PS \div 3 = 42 \div 3 = 14$$

Add 14 to –12 to get Q, and then add 14 more to get R:

$$-12 + 14 = 2 \rightarrow Q$$
$$2 + 14 = 16 \rightarrow R$$

(8) For ∠STU to be trisected, all three angles must be equal. So first set any two angles equal to each other and solve for x. (Any two work, but I use ∠1 and ∠2.)

$$m\angle 1 = m\angle 2$$
$$4x = x + 9$$
$$3x = 9$$
$$x = 3$$

Plugging $x = 3$ into the measure of ∠1 or ∠2 determines both of their measures, because you can assume that they're congruent:

$$m\angle 1 = 4x$$
$$= 4 \cdot 3$$
$$= 12$$

Thus, if $\angle 1$ and $\angle 2$ are congruent, they're both 12°. Finally, check whether $\angle 3$ is also 12° when $x = 3$:

$$m\angle 3 = 5x - 7$$
$$= 5 \cdot 3 - 7$$
$$= 8$$

Nope. Thus, $\angle STU$ is not trisected.

(9) The three angles are in the ratio $4:5:6$, so you first set their measures equal to $4x$, $5x$, and $6x$. Together, the three angles make up a right angle, so set their sum equal to 90° and solve:

$$4x + 5x + 6x = 90$$
$$15x = 90$$
$$x = 6$$

Use $x = 6$ to determine the measure of $\angle PNO$:

$$m\angle PNO = m\angle PNQ + m\angle QNO$$
$$= 5x + 6x$$
$$= 11x$$
$$= 11 \cdot 6$$
$$= 66$$

Of course, you could use $x = 6$ to determine that $\angle PNQ$ is 30° and $\angle QNO$ is 36° and then add them to get 66°.

(*10) Check out the answers:

a. Because $\overline{AC}$ is trisected, AD must equal DE. So set them equal to each other, solve for x, and then plug the answer in to get AD and DE:

$$3x = 5x - 14$$
$$-2x = -14$$
$$x = 7$$

Therefore, $AD = 3x = 3 \cdot 7 = 21$. DE is also, of course, 21, and because $\overline{AC}$ is trisected, EC is also 21. $DC = DE + EC$, so $DC = 42$.

b. No, you can't conclude that $\angle 1 \cong \angle 2$ or that $\angle 1 \cong \angle 3$. Despite the fact that we typically think of rays as angle bisectors or trisectors, the given in this problem is that $\overline{BD}$ and $\overline{BE}$ trisect a *segment*, $\overline{AC}$. This statement tells you only the location of points D and E; it tells you nothing about how the rays divide up $\angle ABC$. $\angle ABC$ might look trisected, but you can't conclude that it is. As it turns out, it's impossible for $\angle ABC$ to be trisected given that $\overline{AC}$ is trisected. $\angle 1$ would be congruent to $\angle 2$ only if $\triangle ABE$ were isosceles (which you can't conclude). And $\angle 1$ would be congruent to $\angle 3$ only if $\triangle ABC$ were isosceles (which you also can't conclude despite the fact that it looks like it is).

(11) And here's another fine solution:

a. Because of the given perpendicularity, you know that $\angle CFE$ and $\angle BFD$ are both 90° angles. Now, $\angle CFD$ and $\angle DFE$ have to add up to 90°, right? So, because $\angle DFE$ is 25°, $\angle CFD$ must be 90° − 25°, or 65°. Then, using the same logic, $\angle BFC$ must be 90° − 65°, which is 25°.

b. The sides of $\angle BFC$ are rays $\overrightarrow{FB}$ and $\overrightarrow{FC}$. If you said segment $\overline{FB}$, you're close. Angles go on forever, and their sides are rays that go on forever. Whether or not you can actually see the rays in the figure is irrelevant.

(12) Because $\overrightarrow{AD} \perp \overrightarrow{BE}$, you know that the measures of both $\angle BGD$ and $\angle AGE$ are 90°. You see that the measure of $\angle BGD = \angle BGC + \angle DGC$ (which is 10°). Thus, $\angle BGC = 90° − 10° = 80°$. Then, because $\angle BGC$ (80°) is four times as big as $\angle AGF$, $\angle AGF = 20°$. Finally, $\angle FGE = 90° − 20°$, which is 70°.

(13) The given perpendicularity tells you that the four big angles are each 90°. (This loose, nontechnical use of "big" may get me pulled over by the math police; don't try it with your geometry teacher.) Because $\overrightarrow{IG}$ bisects right $\angle HIF$, $\angle GIF$ must be 45°. $\angle EIF$ measures 90°, so add these two up to get 135° for $\angle EIG$. Straight $\angle BIG$ (another "big" angle — don't you just love these geometry puns?) is, of course, 180°, so $\angle BIE$ must be 180° − 135°, or 45°. Now, trisect that 45° to get 15° for the three small angles. And finally, two of these 15° angles make up $\angle BID$, so $\angle BID$ measures 30°.

(14) Here's how you do this gnarly problem:

a. The three given pairs of perpendicular segments tell you, by the definition of perpendicular, that the following are right angles: $\angle R$, $\angle G$, $\angle RLN$, and $\angle YLN$. Despite the fact that $\angle Y$ and $\angle A$ look like right angles, you can't conclude that. (But you can conclude that $\angle ANL$ and $\angle GNL$ are right angles — see part b.)

b. Yes, you can conclude that $\angle ANL$ is a right angle, though I haven't covered the necessary concepts yet. What? Is it against the law for me to challenge you with a problem before I've presented the relevant ideas? Well, excuse me! Really, though, you probably could've reached this conclusion if you're familiar with rectangles. Because $\angle R$, $\angle G$, and $\angle RLN$ are right angles, the fourth angle in quadrilateral RGNL, $\angle GNL$, must also be a right angle; that's because the angles in a quadrilateral have to add up to 360°. (The official explanation of the sum of angles in a polygon is in Chapter 6.) Because $\angle GNL$ is a right angle, the angle's supplement, $\angle ANL$, must also be a right angle.

c. $\angle GRY \cap \overleftrightarrow{YL} = \overline{RY}$

d. $\angle GRY \cap \overrightarrow{LY} = \overrightarrow{LY}$

(15) $\angle 1$ and $\angle 5$ are supplementary; $\angle 1$ is 25°, so $\angle 5$ is 155°. Then $\angle 5$ and $\angle 4$ work the same way, so $\angle 4$ is 25°. Because $\angle 2$ is 90°, $\angle 3$ and $\angle 4$ together have to make up another 90° (because $\angle 2$, $\angle 3$, and $\angle 4$ add up to a straight angle, or 180°). Thus, because $\angle 4$ is 25°, $\angle 3$ is 65°. $\angle 4$ and $\angle 6$ are complementary, so $\angle 6$ is also 65°. Finally, going clockwise around the point to $\angle 7$, $\angle 8$, and $\angle 9$, each adjacent pair of angles is supplementary, so $\angle 7$ is 115°, $\angle 8$ is 65°, and $\angle 9$ is 115°.

(16) *Tip:* You can often come up with the correct equation for a word problem like this one by reading through the sentence and translating each word or phrase into its mathematical equivalent.

For this problem, first let x equal the measure of the angle you're trying to find. Then, because you obtain any angle's complement by subtracting the angle's measure from 90° and obtain any angle's supplement by subtracting the angle's measure from 180°, the measure of the complement of the unknown angle is $90 - x$, and the measure of its supplement is $180 - x$. Now you can do the translation:

$$\underbrace{\text{The supplement of an angle}}_{180 - x} \quad \underbrace{\text{is}}_{=} \quad \underbrace{\text{20 greater than}}_{20 +} \quad \underbrace{\text{twice}}_{2 \cdot} \quad \underbrace{\text{the angle's complement.}}_{90 - x}$$

Write this problem like an ordinary equation and solve for x. (But first note that in the following equation, I move the "20 +" to the end of the equation, where it becomes "+ 20." Adding the 20 at the end is more natural. Say you hear someone say, "That's twenty greater than one hundred forty-five." You think $145 + 20$, not $20 + 145$, right? Either works, of course, but now consider the expression, "twenty less than one hundred forty-five." For that, you *have to* subtract the 20 from the 145, not the other way around. Being consistent and putting the 20 at the end is best, regardless of whether you're adding or subtracting.)

Finish the problem:

$$180 - x = 2(90 - x) + 20$$
$$180 - x = 180 - 2x + 20$$
$$x = 20$$

(17) As Sherlock Holmes says in *The Adventure of the Beryl Coronet*, "When you have excluded the impossible, whatever remains, however improbable, must be the truth." So go on and solve this problem just like the great detective would:

a. Yes, it's possible:

$$-3x = -4x - 10$$
$$x = -10$$

Plug $x = -10$ into the angles, and you see that each angle is 30°.

b. Not possible:

$$5x = 3x - 10$$
$$2x = -10$$
$$x = -5$$

Plug $x = -5$ into the angles, and you get negative measures for each angle, which is impossible.

18 Set the vertical angles equal to each other and solve for x:

$$x^2 - 6x = 10 - 3x$$
$$x^2 - 3x - 10 = 0$$
$$(x-5)(x+2) = 0$$
$$x = 5 \quad \text{or} \quad x = -2$$

Now plug each of these two solutions into the original angles. The solution $x = 5$ gives you negative angles, so you reject $x = 5$. The solution $x = -2$ gives you angles of 16°. Because $\angle AQB$ and $\angle DQC$ are the supplements of these angles, they each equal 164°.

IN THIS CHAPTER

» Your starter kit of short geometry proofs

» Looking at the right angle and vertical angle theorems

» Scoping out the complementary and supplementary angle theorems

» Using angle and segment arithmetic

» Standing in: Substitution and transitivity

Chapter **3**

Your First Geometry Proofs

I n this chapter, you get your first taste of the meat of this course: geometry proofs. Do the practice problems carefully, and make sure you understand their solutions. Everything in the subsequent chapters builds on the important proof concepts presented here.

Ready to Try Some Proofs?

Proofs can be quite difficult at first, even the fairly short and straightforward ones in this first section. If you feel a bit lost at the beginning, don't sweat it. Go over the example proof and practice proofs in this section and their solutions as many times as you need for the basic idea of a proof to sink in. And make sure you understand how the flow of logic works (the "bubble-and-arrow" logic I show you in the solutions). If you master the logic and method of doing these first short proofs, you should be able to handle the longer, harder ones later in the book. (If you get stuck, you can check out Chapter 16 for some tips.)

For the proofs in this chapter, you'll need to use several theorems and definitions. I'll give you the theorems as you need them. I gave you the 15 or so definitions you'll need in Chapter 2, so you may want to go back and review them. You might also want to go back to Chapter 1 and reread the section, "What's a Geometry Proof?" Look again at the Fifi the Labradoodle proof and how the bubble-and-arrow logic works. Lastly, here again is the critically important concept about the structure of proofs that appears in that same section of Chapter 1. I'm repeating it verbatim. There may be nothing as important as this idea for understanding how proofs work.

REMEMBER

In a two-column proof, **the idea or ideas in the** *if* **part of each reason must come from the statement column somewhere** *above* **the reason; and the single idea in the** *then* **part of the reason must match the idea in the statement on** *the same line* **as the reason.**

Proofs Involving Complementary and Supplementary Angles

These short proofs involve the simple ideas of complementary angles (two angles that add up to 90°) and supplementary angles (two angles that add up to 180°). But before I give you the complementary and supplementary angle theorems, here are two very simple theorems you'll need later in this chapter and for the rest of the book.

THEOREMS & POSTULATES

All right angles are congruent: If two angles are right angles, then they are *congruent* (they have the same number of degrees).

Many geometry theorems are statements of obvious things. You'll see more of them later in this chapter. But this one about congruent right angles takes the cake in the *well-duh* category. (Put this theorem in your back pocket; you'll use it soon but not in this section.)

THEOREMS & POSTULATES

Vertical angles are congruent: If two angles are vertical angles, then they are congruent. (I'm sure you remember vertical angles from Chapter 2: They're angles across from each other when two lines cross to form an X.) You'll use this theorem later in this chapter, but not in this section.

And now for the complementary and supplementary angle theorems.

THEOREMS & POSTULATES

Here are four easy theorems about pairs of angles that add up to either 90° or 180°.

>> **Complements of the same angle are congruent:** If two angles are each complementary to a third angle, then they're congruent to each other (you have three total angles here).

For example, say you have a 70° angle, $\angle C$. If $\angle A$ is complementary to $\angle C$ and $\angle B$ is also complementary to $\angle C$, then $\angle A \cong \angle B$ (both have to be 20°, right?). Like so many theorems, the idea behind this one is a totally *well-duh* concept.

>> **Complements of congruent angles are congruent:** If two angles are complementary to two other congruent angles, then they're congruent (you're working with four total angles).

For example, if $\angle B \cong \angle C$ (say they're both 40°), and $\angle A$ is complementary to $\angle B$ and $\angle D$ is complementary to $\angle C$, then $\angle A \cong \angle D$ (both have to be 50°).

>> **Supplements of the same angle are congruent:** If two angles are each supplementary to a third angle, then they're congruent to each other (three total angles are involved). This theorem works exactly like the first theorem in this list.

>> **Supplements of congruent angles are congruent:** If two angles are supplementary to two other congruent angles, then they're congruent (this theorem uses four total angles). This theorem works exactly like the second theorem.

TIP

Three or Four Things? Several theorems (the four preceding and many you'll see later) involve either three or four segments or angles. So, when doing a proof, pay attention to whether the proof diagram involves three or four segments or angles. Doing so can help you select the appropriate theorem.

EXAMPLE

Q. Given: $\overline{KS} \perp \overline{SY}$

$\overline{YU} \perp \overline{UK}$

$\angle RST \cong \angle TUR$

Prove: $\angle KSR \cong \angle YUT$

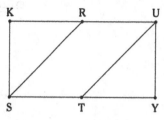

A. *Tip:* Before trying to write down the formal statements and reasons in a two-column proof, it's often a good idea to think through the proof using your own common sense. In other words, try to see why the *prove* statement is true without worrying about how to prove it or worrying about which theorems to use. When you can see why the *prove* statement has to be true, all that remains to be done is to translate your Joe/Jane-six-pack argument into the formal language of a proof.

For example, in this proof, you might say to yourself, "Can I see why angle *KSR* should be congruent to angle *YUT?*" And you could respond, "Sure. Because the segments are perpendicular, angles *KSY* and *YUK* are 90°, and because angle *RST* is congruent to angle *TUR* (say they're both 50°), angle *KSR* has to equal angle *YUT* (they'd both have to equal 40°). Bingo." If you can understand the proof in this commonsense way, then all you have to do is put formalwear on this casual line of reasoning.

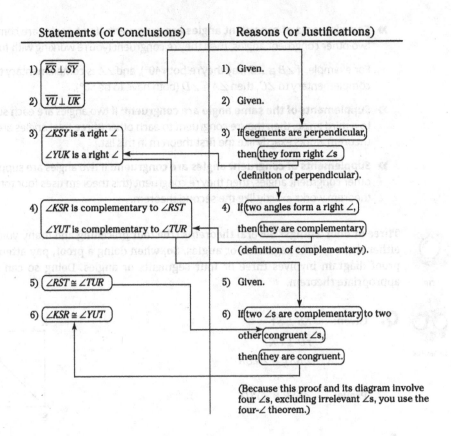

Statements (or Conclusions)	Reasons (or Justifications)
1) $\overline{KS} \perp \overline{SY}$	1) Given.
2) $\overline{YU} \perp \overline{UK}$	2) Given.
3) $\angle KSY$ is a right $\angle$ $\angle YUK$ is a right $\angle$	3) If segments are perpendicular, then they form right $\angle$s (definition of perpendicular).
4) $\angle KSR$ is complementary to $\angle RST$ $\angle YUT$ is complementary to $\angle TUR$	4) If two angles form a right $\angle$, then they are complementary (definition of complementary).
5) $\angle RST \cong \angle TUR$	5) Given.
6) $\angle KSR \cong \angle YUT$	6) If two $\angle$s are complementary to two other congruent $\angle$s, then they are congruent.

(Because this proof and its diagram involve four $\angle$s, excluding irrelevant $\angle$s, you use the four-$\angle$ theorem.)

---

1 Given: $\angle 1 \cong \angle 4$

Prove: $\angle 2 \cong \angle 3$

Note: For this and all proof problems, you should *not* assume that the number of blank lines is the same as the number of steps needed for the proof.

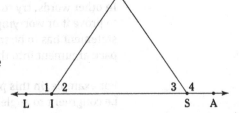

Statements	Reasons

Given: $\overline{ST} \perp \overline{SA}$

$\overline{SR} \perp \overline{SB}$

Prove: $\angle TSR \cong \angle BSA$

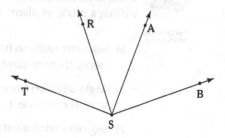

Hint: If you get stuck, go to the solution and copy only Statement 5 and Reason 3 onto this page and then try the proof again.

Statements	Reasons

Proofs Involving Adding and Subtracting Segments and Angles

You get eight more theorems in this section — all of them based on incredibly simple ideas. But despite the fact that the ideas are simple, having to memorize all this mumbo-jumbo lingo may still seem like a pain. If so, I have a tip for you.

TIP

Focus on the ideas behind the theorems. Doing so can help you remember how they're worded. And here's another benefit: If you're doing a proof on a quiz or test and you can't remember exactly how to write some theorem, you can just write the idea of the theorem in your own words. If you get the idea right, you may get partial or even full credit, depending on your teacher's grading style. (And if you're just doing the proof for fun — and who wouldn't? — you can get through the proof using some of your own words. After you're done, you can look up the proper wording of the theorem or theorems you couldn't remember.)

For example, one of the following theorems is based on the incredibly simple notion that if you take two sticks of equal length (say 3 inches and 3 inches) and add them end-to-end to two other equal sticks (say 5 inches and 5 inches), you end up with two equal totals (8 inches and 8 inches, of course). If you understand that idea, you've got the theorem in the bag. Adding equal things to equal things produces equal totals.

THEOREMS & POSTULATES

Without further ado, here are four theorems to use when adding line segments or angles (when writing a proof, students sometimes abbreviate these theorems as "addition").

>> **Segment addition (three total segments):** If a segment is added to two congruent segments, then the sums are congruent.

>> **Angle addition (three total angles):** If an angle is added to two congruent angles, then the sums are congruent.

>> **Segment addition (four total segments):** If two congruent segments are added to two other congruent segments, then the sums are congruent.

>> **Angle addition (four total angles):** If two congruent angles are added to two other congruent angles, then the sums are congruent.

THEOREMS & POSTULATES

If you're subtracting segments or angles, here are four more theorems to choose from (after you get a handle on these theorems, you may simply write "subtraction").

>> **Segment subtraction (three total segments):** If a segment is subtracted from two congruent segments, then the differences are congruent.

>> **Angle subtraction (three total angles):** If an angle is subtracted from two congruent angles, then the differences are congruent.

>> **Segment subtraction (four total segments):** If two congruent segments are subtracted from two other congruent segments, then the differences are congruent.

>> **Angle subtraction (four total angles):** If two congruent angles are subtracted from two other congruent angles, then the differences are congruent.

TIP

Here are a couple huge tips that you can use when working on any proof. You can see them in action in the first example in this section.

>> **Use every given.** You have to do something with every given in a proof. So, if you're not sure how to do a proof, don't give up until you've at least asked yourself, "Why did they give me this given? And why did they give me that given?" If you then write down what follows from each given (even if you don't know how that information can help you), you might see how to proceed. You may have a geometry teacher (or mathematician friend) who likes to throw you the occasional curveball, but in every geometry text that I know, the authors don't give you irrelevant givens. And that means that *every given is a built-in hint.*

>> **Work backwards.** Thinking about how a proof will end — what the last and second-to-last lines will look like — is often very helpful. In some proofs, you may be able to work backwards from the final statement to the second-to-last statement and then to the third-to-last statement and maybe even to the fourth-to-last. Doing proofs this way is a little like doing one of those mazes you see in a newspaper or magazine. You can begin by working on a path from the Start point. Then, if you get stuck, you can work on a path from the Finish point, taking that as far as you can. And then you can go back to where you left off and try to connect the ends of the two paths.

Q. Given: *I* is the midpoint of $\overline{RN}$

EXAMPLE

   R and *N* trisect $\overline{GD}$

Prove: *I* is the midpoint of $\overline{GD}$

G   R   I   N   D

A. Use every given. In this example proof, pretend that you have no idea how to begin. Just do something with the two givens. Ask yourself why someone would tell you about a midpoint. Well, because that tells you that you have two congruent segments, of course. And why would someone give you the trisection points? Because that given tells you that you have three congruent segments (though you use only two of them).

Statements	Reasons
1) *I* is the midpoint of $\overline{RN}$	1) Given.
2) *R* and *N* trisect $\overline{GD}$	2) Given.
3) $\overline{RI} \cong \overline{IN}$	3) If a point is the midpoint of a segment, then it divides it into two congruent segments.
4) $\overline{GR} \cong \overline{ND}$	4) If two points trisect a segment, then they divide it into three congruent segments.
5) $\overline{GI} \cong \overline{ID}$	5) If two congruent segments are added to two other congruent segments, then the sums are congruent.
6) *I* is the midpoint of $\overline{GD}$	6) If a point divides a segment into two congruent segments, then it is the midpoint of the segment (reverse of definition of midpoint).

Okay, here's where working backwards can help: Say you can figure out lines 3 and 4 in the preceding proof but aren't sure where to go next. No worries. Jump to the end of the proof. You know the final statement has to be the *prove* statement (*I* is the midpoint of $\overline{GD}$). Now ask yourself what you'd need to know to draw that conclusion. Well, to conclude that a point is a midpoint, you need a segment that's been cut into two congruent segments, right? So, you don't have to be a mathematical genius to see that the second-to-last statement has to be $\overline{GI} \cong \overline{ID}$.

After you see that point, all you have to do is figure out why *that* would be true. So, you then go back to where you left off (line 4), and hopefully you then see that you can add two pairs of congruent segments to get $\overline{GI} \cong \overline{ID}$.

3 Given: $\angle GBU \cong \angle SBM$

Prove: $\angle GBM \cong \angle SBU$

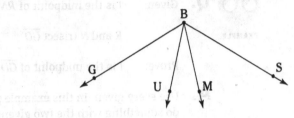

Statements	Reasons

4 Given: R is the midpoint of $\overline{BS}$

U and N trisect $\overline{BS}$

Prove: R is the midpoint of $\overline{UN}$

B   U   R   N   S

Hint: If you have a hard time with this one, take Statements 4 and 6 and Reason 3 from the solutions section and copy them here. But don't do this unless you absolutely have to.

Statements	Reasons

5 Given: $\overline{QY}$ bisects $\angle ZQX$

$\angle ZQW \cong \angle XQJ$

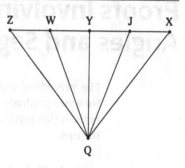

Prove: $\overline{QY}$ bisects $\angle WQJ$

Hint: Don't forget to use all the givens in your proof (you might want to make them your first two steps). If you're really stumped, go to the solution and copy just the *if* part of Reason 3 onto this page.

Statement	Reason

6 You can do the following proof in four different ways, using four different sets of theorems. Don't write out four two-column proofs (unless you feel like it). Just write your game plans for the four alternatives. *Hint:* Two of the versions use vertical angles, two use $\angle ACT$ instead, two use angle subtraction, and two use complementary angles.

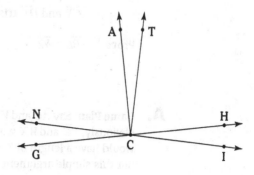

Given: $\overrightarrow{CA} \perp \overrightarrow{GH}$

$\overrightarrow{CT} \perp \overrightarrow{NI}$

Prove: $\angle NCA \cong \angle HCT$ (paragraph proof)

Proofs Involving Multiplying and Dividing Angles and Segments

The preceding section lets you work on addition and subtraction of segments and angles. Now you graduate to multiplication and division of segments and angles. The two new theorems in this section can be a bit tricky to use correctly, so study these proofs carefully and heed the tips.

THEOREMS & POSTULATES

>> **Like Multiples:** If two segments (or angles) are congruent, then their like multiples are congruent. This statement just means that if you have, say, two congruent segments, then 3 times one segment equals 3 times the other, or 4 times one equals 4 times the other — another *well-duh* idea.

>> **Like Divisions:** If two segments (or angles) are congruent, then their like divisions are congruent. All this statement tells you is that if you have, say, two congruent angles, then 1/2 of one equals 1/2 the other, or 1/3 of one equals 1/3 of the other.

TIP

Do you see something twice? If the givens in a proof mention midpoint, bisect, or trisect *twice* (or something else that amounts to the same thing), then there's a pretty good chance that you'll want to use the Like Multiples Theorem or the Like Divisions Theorem in the proof.

Notice that this tip applies to both example problems and the three practice problems.

EXAMPLE

Q. Given: $\overline{AC} \cong \overline{VX}$

$\overline{AB} \cong \overline{VW}$

$\overline{CX}$ and $\overline{DY}$ trisect both $\overline{BE}$ and $\overline{WZ}$

Prove: $\overline{BE} \cong \overline{WZ}$

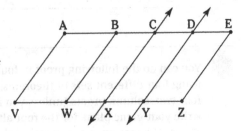

A. **Game Plan:** Say $\overline{AC}$ and $\overline{VX}$ both had a length of 10, and $\overline{AB}$ and $\overline{VW}$ were both 6. Then, obviously, $\overline{BC}$ and $\overline{WX}$ would both be 4. Then, since $\overline{BE}$ and $\overline{WZ}$ are both trisected, each would have a length of $3 \cdot 4$, or 12. That's all there is to this proof. And here's how you write out this simple argument in the formal way:

Statements	Reasons
1) $\overline{AC} \cong \overline{VX}$	1) Given.
2) $\overline{AB} \cong \overline{VW}$	2) Given.
3) $\overline{BC} \cong \overline{WX}$	3) If two congruent segments are subtracted from two other congruent segments, then the differences are congruent (segment subtraction; four-segment version).
4) $\overline{CX}$ and $\overline{DY}$ trisect both $\overline{BC}$ and $\overline{WZ}$	4) Given.
5) $\overline{BE} \cong \overline{WZ}$	5) If segments are congruent, then their like multiples are congruent.

TIP

When, like in the proof here, **you go from a statement about small things** (like $\overline{BC}$ and $\overline{WX}$) **to a statement about big things** (like $\overline{BE}$ and $\overline{WZ}$), **use the Like *Multiples* Theorem.**

TIP

It can be *very* helpful to make up lengths of segments (or sizes of angles) like I just did in the game plan of this example proof. Look back to where I said, "Say $\overline{AC}$ and $\overline{VX}$ both had a length of 10, and $\overline{AB}$ and $\overline{VW}$ were both 6." You don't know the lengths of those four segments, but making up their lengths like this will often help you see the logic of the proof. After making up sizes of things, you can see how the proof works by doing some simple arithmetic (like I did when I determined that $\overline{BC}$ and $\overline{WX}$ would both be 4 and that, therefore, $\overline{BE}$ and $\overline{WZ}$ would both be $3 \cdot 4$, or 12). When you use this strategy, **you can make up lengths for segments (and sizes for angles) that are listed in the givens and sometimes for *unnamed* segments and angles. But DO NOT make up lengths for segments and sizes for angles listed in the prove statement.** (Note that in the game plan for this example proof, I ended up *concluding* that $\overline{BE}$ and $\overline{WZ}$ were both 12, but that was the result of the simple arithmetic I did. I did not start by making up their lengths.)

EXAMPLE

Q. Given: $\angle 1 \cong \angle 2$

$\overrightarrow{QU}$ bisects $\angle RQS$

$\overrightarrow{ST}$ bisects $\angle RSQ$

Prove: $\angle RQU \cong \angle RST$

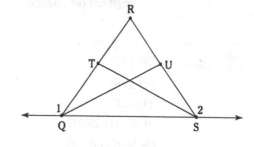

A. **Game Plan:** Okay. You have ∠1 equal to ∠2 (say they're both 120°). Their supplements would then have to be equal (they'd both be 60°). Each of those is bisected, so ∠RQU and ∠RST would both be 30°. Piece of cake.

Statements	Reasons
1) ∠1 ≅ ∠2	1) Given.
2) ∠RQS is supplementary to ∠1 ∠RSQ is supplementary to ∠2	2) If two angles form a straight angle (assumed from diagram), then they are supplementary (reverse of definition of supplementary).
3) ∠RQS ≅ ∠RSQ	3) If two angles are supplementary to two other congruent angles, then they are congruent (supplements of congruent angles) (Statements 1 and 2).
4) $\overline{QU}$ bisects ∠RQS	4) Given.
5) $\overline{ST}$ bisects ∠RSQ	5) Given.
6) ∠RQU ≅ ∠RST	6) If angles are congruent, then their like divisions are congruent (Like Divisions) (Statements 3, 4, and 5).

When, like in this last proof, **you go from a statement about big things** (∠RQS and ∠RSQ) **to a statement about small things** (like ∠RQU and ∠RST), **you use the Like Divisions Theorem.**

When you're new to proofs, it's easy to get confused about when to use the definitions of midpoint, bisect, or trisect and when to use the Like Divisions Theorem. So, take heed: **Use the definitions when you want to show that two or three parts of the *same*** **segment or *same* angle are equal to each other. Use Like Divisions, in contrast, when** **you want to show that a part of one segment (or angle) is equal to a part of a *different*** **segment (or angle).**

 7 Given: $\overline{NO} \perp \overline{NI}$

$\overline{NO} \perp \overline{OE}$

∠1 ≅ ∠2

$\overline{NI}$ bisects ∠DNG

$\overline{OE}$ bisects ∠TOG

Prove: ∠DNG ≅ ∠TOG

Hint: Want a little help? Check out Statements 1, 2, and 3 on the solution page.

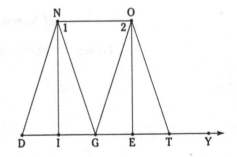

Statements	Reasons

8 Given: $\overline{SD} \cong \overline{UE}$

M is the midpoint of $\overline{SU}$

G is the midpoint of $\overline{DE}$

Prove: $\overline{SM} \cong \overline{GE}$

Statements	Reasons

9 Given: $\overline{EA} \perp \overline{ED}$

$\overrightarrow{VW} \perp \overrightarrow{VZ}$

$\overrightarrow{EB}$ and $\overrightarrow{EC}$ trisect $\angle AED$

$\overrightarrow{VX}$ and $\overrightarrow{VY}$ trisect $\angle WVZ$

Prove: $\angle AEV \cong \angle WVE$

Hint: If this problem seems a bit tough, copy only
Statement 9 and Reason 9 from the solution and
try to work backwards from there.

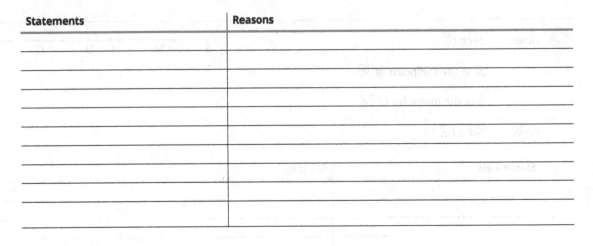

Statements	Reasons

Proofs Involving the Transitive and Substitution Properties

THEOREMS & POSTULATES

The transitive and substitution properties should be familiar to you from algebra. You may have used the idea of transitivity in this way: If $a = b$ and $b = c$, then $a = c$; or if $a > b$ and $b > c$, then $a > c$. Transitivity works the same in geometry: You use it like with those algebra examples but to show congruence instead of equality (you almost never, however, use the inequality version). And you've certainly used substitution in algebra — like if $x = 2y - 5$ and $4x - 3y = 10$, you can switch the x with the $2y - 5$ (because they're equal, of course) and write $4(2y - 5) - 3y = 10$. This property works the same in geometry: When two objects are congruent, you can switch 'em.

>> **Transitive Property (for three segments or angles):** If two segments (or angles) are each congruent to a third segment (or angle), then they're congruent to each other. For example, if $\angle A \cong \angle B$ and $\angle B \cong \angle C$, then $\angle A \cong \angle C$ ($\angle A$ and $\angle C$ are each congruent to $\angle B$, so they're congruent to each other).

» **Transitive Property (for four segments or angles):** If two segments (or angles) are congruent to congruent segments (or angles), then they're congruent to each other. For example, if $\overline{AB} \cong \overline{CD}$, $\overline{CD} \cong \overline{EF}$, and $\overline{EF} \cong \overline{GH}$, then $\overline{AB} \cong \overline{GH}$. ($\overline{AB}$ and $\overline{GH}$ are congruent to the congruent segments $\overline{CD}$ and $\overline{EF}$, so they're congruent to each other.)

» **Substitution Property:** If two geometric objects (segments, angles, triangles, and so on) are congruent and you have a statement involving one of them, you can pull the switcheroo and replace the one with the other. For example, if $\angle A \cong \angle B$ and $\angle B$ is supplementary to $\angle C$, then $\angle A$ is supplementary to $\angle C$.

TIP

You use the Transitive Property as the reason when the statement says things are congruent; you use the Substitution Property for the reason when the statement says anything else.

TIP

And one more thing: You'll be less likely to mix up substitution with other theorems if you note that **like with transitivity, other theorems** (addition, subtraction, complements and supplements of congruent angles, and so on) **go with statements about congruent things; substitution does not.**

Q. Given: $\angle 2 \cong \angle 3$

Prove: $\angle 1 \cong \angle 3$

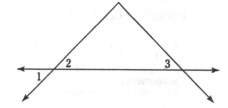

A. Here's how this one unfolds:

Statements	Reasons
1) $\angle 2 \cong \angle 3$	1) Given.
2) $\angle 1 \cong \angle 2$	2) Vertical angles are congruent.
3) $\angle 1 \cong \angle 3$	3) If two angles are each congruent to a third angle, then they are congruent to each other (Transitive Property).

Did it occur to you that you could use substitution instead of transitivity for Reason 3? That's correct — sort of. You could use substitution in Step 3 because you can essentially put $\angle 3$ where $\angle 2$ is. The switch works this way because transitivity is a special case of substitution. However, you probably want to use the property as I do (rebels excepted), because that's probably what your geometry teacher and mathematician buddies want. (For info on how to keep the properties straight, see the preceding tips in this section.)

Q. Given: $\angle 2 \cong \angle 3$

EXAMPLE

Prove: $\angle 1$ is supplementary to $\angle 3$

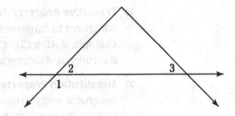

A. The proof, dear reader:

Statements	Reasons
1) $\angle 1$ is supplementary to $\angle 2$	1) If two angles form a straight angle (assumed from diagram), then they are supplementary.
2) $\angle 2 \cong \angle 3$	2) Given.
3) $\angle 1$ is supplementary to $\angle 3$	3) Substitution (putting $\angle 3$ where $\angle 2$ was).

---

10 Given: $\overrightarrow{AC}$ bisects $\angle BAD$

Prove: $\angle 1 \cong \angle 3$

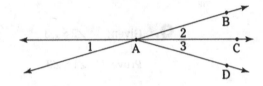

Statements	Reasons

---

11 Given: $\overrightarrow{MB}$ bisects $\angle AMC$

$\overrightarrow{MC}$ bisects $\angle BMD$

Prove: $\angle 4 \cong \angle 6$

Hint: If you get stuck, copy Statements 1 and 3 and Reason 5 from the solution page and then take it from the top.

Statements	Reasons

12 Given: $\overline{TO} \perp \overline{GO}$

Prove: ∠1 is complementary to ∠2

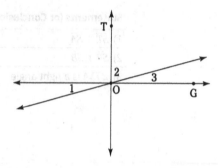

Statements	Reasons

Solutions

1

Statements (or Conclusions)	Reasons (or Justifications)
1) ∠1 ≅ ∠4	1) Given.
2) ∠LIS is a straight angle ∠ASI is a straight angle	2) Assumed from diagram. (The vast majority of reasons you'll use in proofs will come from your handy lists of definitions, theorems, postulates, and properties. This is one of the few odd exceptions. Some geometry teachers let you skip this step and go right to Step 3.)
3) ∠2 is supplementary to ∠1 ∠3 is supplementary to ∠4	3) If two angles form a straight angle, then they are supplementary (definition of supplementary).
4) ∠2 ≅ ∠3	4) If two angles are supplementary to two other congruent angles, then they are congruent.

2

Statements (or Conclusions)	Reasons (or Justifications)
1) $\overline{ST} \perp \overline{SA}$	1) Given.
2) $\overline{SR} \perp \overline{SB}$	2) Given.
3) ∠TSA is a right angle	3) If two rays are perpendicular, then they form a right angle. (If you understand the if-then rule for reasons that I explain in Chapter 1, then this reason just about writes itself. The *if* part of this reason must come from a statement above it, namely Statement 1 or 2. The only fact in those statements concerns perpendicularity. So basically, this reason has to begin with "If perpendicular". And the only thing that can follow "If perpendicular," is "then right angle.")
4) ∠BSR is a right angle	4) Same as Reason 3.
5) ∠TSR is complementary to ∠RSA ∠BSA is complementary to ∠RSA	5) If two angles form a right angle, then they are complementary (definition of complementary).
6) ∠TSR ≅ ∠BSA	6) If two angles are each complementary to a third angle, then they are congruent to each other. (This proof and its diagram involve three angles, so you use the three-angle theorem.)

3

Statements	Reasons
1) ∠GBU ≅ ∠SBM	1) Given.
2) ∠GBM ≅ ∠SBU	2) If an angle (∠UBM) is added to two congruent angles (∠GBU and ∠SBM), then the sums are congruent (addition of angles; three-angle version).

This proof brings me to my next tip:

TIP

If the angles (or segments) in the *prove* statement are *larger* than the given angles (or segments), the proof may call for one of the *addition* theorems.

Statements	Reasons
1) *R* is the midpoint of $\overline{BS}$	1) Given.
2) *U* and *N* trisect $\overline{BS}$	2) Given.
3) $\overline{BR} \cong \overline{RS}$	3) A midpoint divides a segment into two congruent segments (definition of midpoint).
4) $\overline{BU} \cong \overline{NS}$	4) Trisection points divide a segment into three congruent segments (definition of trisection).
5) $\overline{UR} \cong \overline{RN}$	5) If two congruent segments are subtracted from two other congruent segments, then the differences are congruent (subtraction of segments; four-segment version).
6) *R* is the midpoint of $\overline{UN}$	6) If a point divides a segment into two congruent segments, then it's the midpoint of the segment (reverse of definition of midpoint).

Note that in contrast to the preceding problem, in this proof, the things you're trying to prove something about ($\overline{UR}$ and $\overline{RN}$) are *smaller* than the things in the given ($\overline{BR}$ and $\overline{RS}$ are sort of in the given).

If the segments (or angles) in the *prove* statement are *smaller* than the ones in the given, one of the *subtraction* theorems may be the ticket.

TIP

⑤

Statements	Reasons
1) $\overrightarrow{QY}$ bisects $\angle ZQX$	1) Given.
2) $\angle ZQW \cong \angle XQJ$	2) Given.
3) $\angle ZQY \cong \angle XQY$	3) If a ray bisects an angle, then it divides it into two congruent angles (definition of bisect).
4) $\angle WQY \cong \angle JQY$	4) If two congruent angles are subtracted from two other congruent angles, then the differences are congruent (subtraction of angles; four-angle version).
5) $\overrightarrow{QY}$ bisects $\angle WQJ$	5) If a ray divides an angle into two congruent angles, then the ray bisects the angle (reverse of definition of bisect).

⑥ All four game plans use, of course, the two right angles.

Game Plan 1: You have the two congruent vertical angles. One is the complement of $\angle NCA$; the other is the complement of $\angle HCT$. Therefore, you finish with the complements of congruent angles theorem. (Assuming each statement contains only a single fact, this method takes eight steps. Try it.)

Game Plan 2: This method is the same as Game Plan 1 except that you subtract the congruent vertical angles from the congruent right angles. The final reason is, therefore, the four-angle version of *angle subtraction*. (This strategy takes seven steps. Give it a go.)

Game Plan 3: Use $\angle ACT$. $\angle NCA$ and $\angle HCT$ are both complements of $\angle ACT$. You're done, because complements of the same angle are congruent. (This method also takes seven steps. Go for it.)

Game Plan 4: This time, you just subtract ∠ACT from the two congruent right angles. The final reason is the three-angle version of *angle subtraction*. (The winner! — only six steps.)

*7 **Game Plan:** You have the right angles and ∠1 ≅ ∠2 (say they're both 70°). So, their complements (∠ING and ∠EOG) would both measure 20°. Then because of the bisections, the angles you're trying to prove equal to each other would both be 2 · 20, or 40°. That's it.

Statements	Reasons
1) $\overline{NO} \perp \overline{NI}$ $\overline{NO} \perp \overline{OE}$	1) Given.
2) ∠INO is a right angle ∠EON is a right angle	2) Definition of perpendicular (Statement 1).
3) ∠1 ≅ ∠2	3) Given.
4) ∠ING is complementary to ∠1 ∠EOG is complementary to ∠2	4) Definition of complementary angles (Statement 2).
5) ∠ING ≅ ∠EOG	5) Complements of congruent angles are congruent (Statements 3 and 4).
6) $\overline{NI}$ bisects ∠DNG $\overline{OE}$ bisects ∠TOG	6) Given.
7) ∠DNG ≅ ∠TOG	7) If angles are congruent (∠ING and ∠EOG), then their like multiples are congruent (Statements 5 and 6).

8 **Game Plan:** SD equals UE (say they're both 10). If UD is 2, then both SU and DE would be 8. Then the midpoints cut each of those in half, so that makes SM and GE both 4. Bingo.

Statements	Reasons
1) $\overline{SD} \cong \overline{UE}$	1) Given.
2) $\overline{SU} \cong \overline{DE}$	2) If a segment is subtracted from two congruent segments, then the differences are congruent (segment subtraction; three-segment version) (Statement 1 and diagram).
3) M is the midpoint of $\overline{SU}$	3) Given.
4) G is the midpoint of $\overline{DE}$	4) Given.
5) $\overline{SM} \cong \overline{GE}$	5) If segments are congruent ($\overline{SU}$ and $\overline{DE}$), then their like divisions are congruent (half of one equals half of the other) (Statements 2, 3, and 4).

⑨ **Game Plan:** You have the two 90° angles. Each is trisected, so all the small angles measure 30°. Because ∠BEA and ∠XVW measure 30°, ∠AEV and ∠WVE each have to be 150°. Sweet.

Statements	Reasons
1) $\overline{EA} \perp \overline{ED}$	1) Given.
2) $\overline{VW} \perp \overline{VZ}$	2) Given.
3) ∠AED is a right angle ∠WVZ is a right angle	3) Definition of perpendicular (1, 2).
4) ∠AED ≅ ∠WVZ	4) All right angles are congruent (3).
5) $\overline{EB}$ and $\overline{EC}$ trisect ∠AED	5) Given.
6) $\overline{VX}$ and $\overline{VY}$ trisect ∠WVZ	6) Given.
7) ∠AEB ≅ ∠WVX	7) If angles are congruent (the two right angles), then their like divisions are congruent (a third of one equals a third of the other) (4, 5, 6).
8) ∠AEV is supplementary to ∠AEB ∠WVE is supplementary to ∠WVX	8) If two angles form a straight angle (assumed from diagram), then they are supplementary (definition of supplementary).
9) ∠AEV ≅ ∠WVE	9) Supplements of congruent angles are congruent (7, 8).

⑩

Statements	Reasons
1) $\overline{AC}$ bisects ∠BAD	1) Given.
2) ∠2 ≅ ∠3	2) Definition of bisect.
3) ∠1 ≅ ∠2	3) Vertical angles are congruent.
4) ∠1 ≅ ∠3	4) Transitive Property.

⑪ **Game Plan:** Think backwards — how can you get ∠4 ≅ ∠6? Well, ∠4 and ∠3 are congruent vertical angles, as are ∠6 and ∠1. Thus, if you can get ∠1 ≅ ∠3, you have it. The two bisectors make ∠1 ≅ ∠2 and ∠2 ≅ ∠3. Thus, ∠1 ≅ ∠3 by the Transitive Property. Bingo.

Statements	Reasons
1) $\overline{MB}$ bisects ∠AMC	1) Given.
2) ∠1 ≅ ∠2	2) Definition of bisect.
3) $\overline{MC}$ bisects ∠BMD	3) Given.
4) ∠2 ≅ ∠3	4) Definition of bisect.
5) ∠2 ≅ ∠3	5) Transitive Property (for three angles).
6) ∠1 ≅ ∠6	6) Vertical angles are congruent.
7) ∠3 ≅ ∠4	7) Vertical angles are congruent.
8) ∠4 ≅ ∠6	8) Transitive Property (for four angles). If angles (4 and 6) are congruent to congruent angles (1 and 3), then they (4 and 6) are congruent to each other.

(12) Statements	Reasons
1) $\overline{TO} \perp \overline{GO}$	1) Given.
2) ∠TOG is a right angle	2) Definition of perpendicular.
3) ∠3 is complementary to ∠2	3) Definition of complementary.
4) ∠1 ≅ ∠3	4) Vertical angles are congruent.
5) ∠1 is complementary to ∠2	5) Angle substitution.

Triangles, Proof and Non-Proof Problems

2

Start off your love affair with triangles by working out non-proof problems that cover concepts such as area, altitudes, medians, angle bisectors, perpendicular bisectors, the Pythagorean Theorem, families of right triangles, and more.

Get lots of practice proving that triangles are congruent and then using CPCTC (Congruent Parts of Congruent Triangles are Congruent).

Chapter 4

Triangle Fundamentals and Other Cool Stuff (No Proofs)

There's no upper limit to how many sides a polygon can have, but the lower limit is three — and that makes the triangle sort of a special shape. And for some reason, the number three seems to have a certain universal appeal: the Three Stooges, the Three Wise Men, three blind mice, Goldilocks and the three bears, Three Dog Night, three strikes and you're out, and so on. So, I give you the triangles: three angles, three sides, three medians, three altitudes, three angle bisectors, three perpendicular bisectors, and three "centers" (plus the centroid).

Triangle Types and Triangle Basics

Six basic terms describe different types of triangles. Here's a great way to remember them: A triangle has three sides and three angles. Well, three of the following terms are about sides, and three are about angles.

REMEMBER

Every triangle belongs to one of these three categories about sides.

>> **Scalene:** A scalene triangle has no equal sides.

>> **Isosceles:** An isosceles triangle has at least two equal sides. The two equal sides are called *legs;* the third side is the *base.* The two angles touching the base, called *base angles,* are equal. The angle between the two legs is the *vertex angle.*

>> **Equilateral:** An equilateral triangle has three equal sides (thus, every equilateral triangle is also isosceles). Note that an equilateral triangle is also *equiangular* because it has three equal angles (each is 60°). For polygons with four or more sides, the distinction between *equilateral* and *equiangular* is important. Not so for triangles, because both terms refer to the very same triangle.

REMEMBER

Every triangle also belongs to one of these three groups concerning angles.

>> **Acute:** An acute triangle has three acute angles (angles less than 90°, of course).

>> **Right:** A right triangle has one right angle and two acute angles (the two short sides touching the right angle are the *legs;* the longest side across from the right angle is called the *hypotenuse*).

>> **Obtuse:** An obtuse triangle has a single obtuse angle (more than 90°); the other two angles are acute. And here's one more thing about the angles in a triangle that you may already know:

THEOREMS & POSTULATES

The sum of the measures of the three angles in a triangle is always 180°.

TIP

Whenever possible, don't just memorize math formulas, concepts, and so on as raw facts that can only be learned by rote. Instead, look for some reason why they're true or find a connection between the new idea and something you already know. For instance, to remember the sum of the angles in a triangle, picture the triangle you get when you cut a square in half along its diagonal: You can easily see that the three angles of such a triangle are 45°, 45°, and 90° — which add up to 180°.

EXAMPLE

Q. Classify this triangle as scalene, isosceles, or equilateral.

Hint: Don't forget that the sides of a triangle can't be negative or zero (or imaginary, like $\sqrt{-10}$) and that each pair of sides of a triangle must add up to more than the third side.

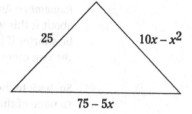

A. This problem isn't simple. Here's how your argument should go: The triangle is scalene unless at least two of the sides are equal. So, try the three different pairs of sides and see what happens if you set them equal to each other.

$$25 = 75 - 5x$$
$$5x = 50$$
$$x = 10$$

No good: Plugging $x = 10$ into $10x - x^2$ gives you a side with a length of zero. Next, if

$$75 - 5x = 10x - x^2$$
$$x^2 - 15x + 75 = 0$$

Now solve for x with the quadratic formula. "What?" you say. "You expect me to remember the quadratic formula?" Yeah, sure, I know this is a geometry book, but I don't think reviewing some algebra as important as the quadratic formula will kill you. Do you have your helmet on?

$$ax^2 + bx + c = 0$$

$$x = \frac{-b \pm \sqrt{b^2 - 4ac}}{2a}$$

$$= \frac{15 \pm \sqrt{(-15)^2 - 4(1)(75)}}{2}$$

$$= \frac{15 \pm \sqrt{225 - 300}}{2}$$

$$= \frac{15 \pm \sqrt{-75}}{2}$$

No good. A negative under the square root means you have no real solutions. Finally, if

$$25 = 10x - x^2$$
$$x^2 - 10x + 25 = 0$$
$$(x - 5)(x - 5) = 0$$
$$x = 5$$

Plugging 5 into the third side $(75 - 5x)$ gives you $75 - 5 \cdot 5$, or 50, so when $x = 5$, you might think the three sides could be 25, 25, and 50. "But wait!" you should say. "No triangle can have sides of 25, 25, and 50!"

Remember: Any two sides of a triangle must add up to *more than* the third side. Think about it this way: If you walk, say, from vertex A to vertex B in a triangle, the trip has to be shorter if you walk straight along $\overline{AB}$ than if you go out of your way and walk along the two other sides.

So, back to the problem. Setting the three pairs of sides equal to each other doesn't work, so none of the sides are equal. Therefore, the triangle is scalene. And that's a wrap.

---

1 Classify these triangles as scalene, isosceles, or equilateral.

a)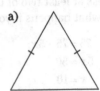

b)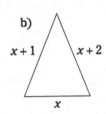

---

2 If △ISO is isosceles and its perimeter is more than 10, which side is the base, and how long are the three sides?

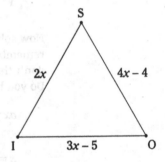

---

3 The angles of a triangle are in the ratio of 4 : 5 : 6. Is it an acute, right, or obtuse triangle? Is it scalene, isosceles, or equilateral?

---

4 Classify the following triangles as acute, obtuse, or right.

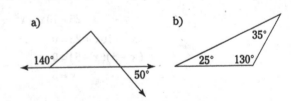

a)

b)

5 Are the following statements true *always*, *sometimes*, or *never*?

a. An equilateral triangle is isosceles.

b. An isosceles triangle is equilateral.

c. A right triangle is isosceles.

d. If two of the angles in a triangle are 70° and 55°, the triangle is isosceles.

e. The base angles of an obtuse isosceles triangle are each 40°.

f. The base angles of an acute isosceles triangle are each 40°.

g. Two of the angles in an obtuse triangle are supplementary (add up to 180°).

h. Two of the angles in an acute triangle are complementary (add up to 90°).

Altitudes, Area, and the Super Hero Formula

In this section, I cover some concepts that you've probably known for a long time, like how to find the area or the height of a triangle. But you will find here some other ideas that I bet you don't know, like why Hero is a real geometry superhero (and that there's a second way to find a triangle's area).

REMEMBER

First, take a look at **the formula for the area of triangle:**

$$\text{Area}_\Delta = \frac{1}{2}\text{base} \cdot \text{height}$$

Area, of course, is usually measured in some kind of units^2, like square feet, square meters, or square centimeters.

TIP

A triangle's height is the distance from its peak straight down. The *height*, or *altitude*, of a triangle is just what you'd expect it to be — you know, its height. Think of altitude this way: If you have an actual, physical triangle — say, cut out of cardboard — and you stand it up on a table, its height or altitude is the distance from its peak straight down to the table. Check out the two triangles in Figure 4-1.

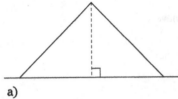
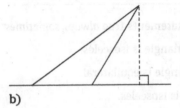

a)   b)

REMEMBER

You can stand a triangle up three different ways depending on which side you put flat on the table, so every triangle has three separate altitudes. Depending on which type of triangle you have, the altitudes can have the same or different lengths.

>> **Scalene triangles:** The three altitudes have different lengths.

>> **Isosceles triangles:** Two of the altitudes have the same length.

>> **Equilateral triangles:** All three altitudes have the same length.

Also, as you can see in Figure 4-1b, sometimes an altitude is outside the triangle. This situation occurs when the triangle is obtuse. Two of the three altitudes in every obtuse triangle are outside the triangle; the third altitude is inside the triangle. And for every right triangle, the two legs are also altitudes, and the third altitude is inside the triangle. All three altitudes of an acute triangle are inside the triangle.

REMEMBER

Alternate triangle area formula. The most common way of figuring a triangle's area is by plugging the triangle's base and height into the regular area formula. But if all you know are the triangle's three sides, you can use the following nifty alternate formula attributed to Hero of Alexandria (who lived from 10 to 70 AD — or CE if you prefer).

$$\text{Area}_\Delta = \sqrt{S(S-a)(S-b)(S-c)}$$

In this formula, a, b, and c are the length of the triangle's sides, and S is the triangle's semiperimeter (half the perimeter: $S = \frac{a+b+c}{2}$).

REMEMBER

And here's one more triangle area formula for you:

The area of an equilateral triangle with side s is $\frac{s^2\sqrt{3}}{4}$.

Q. Given:

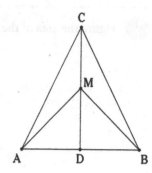

EXAMPLE

$AB = 10$

D is the midpoint of $\overline{AB}$

M is the midpoint of $\overline{CD}$

$\overline{AB} \perp \overline{CD}$

$\triangle BDM$ is isosceles

Find: Area of $\triangle ACM$

A. To use the ordinary area formula, you need a base and a height of $\triangle ACM$. A base and its corresponding height are always perpendicular, so the 90° angle at D is the place to look. Take this book and rotate it 90° clockwise. Now, picture $\triangle ACM$ standing up on a table, where the tabletop runs along $\overline{CD}$. Side $\overline{CM}$ is on the table, so that's the base. And the height goes from the peak (A) straight down to the table at D, so the height is $\overline{AD}$. To use the area formula, you need the lengths of $\overline{CM}$ and $\overline{AD}$.

$AB = 10$ and D is the midpoint of $\overline{AB}$, so $AD = 5$. One down, one to go. $\triangle BDM$ is isosceles, so two of its sides are equal. It's also a right triangle with hypotenuse $\overline{MB}$, so the two equal sides have to be $\overline{DM}$ and $\overline{DB}$ (the hypotenuse is always longer than the legs). DB equals AD, which is 5, so DB is 5; thus, so is DM. Because M is the midpoint of $\overline{CD}$, CM is 5 as well. So, the base and height are both 5.

Now just use the formula:

$$\text{Area}_{\triangle ACM} = \frac{1}{2} \text{ base} \cdot \text{height}$$
$$= \frac{1}{2} \cdot 5 \cdot 5$$
$$= 12.5$$

The area is 12.5 units^2.

 Recalling that some altitudes may be outside the triangle (like in Figure 4-1b shown earlier in the chapter), draw in the three altitudes of the following triangle.

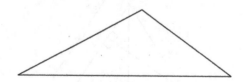

7 Figure the area of the big triangle in four different ways.

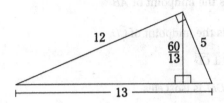

8 Compute the area of rectangle *ABDE* and then the areas of △*ACE*, △*AGE*, and △*APE*. What two conclusions can you draw about these areas?

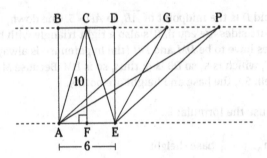

9 Given: $MT = 6$

Find: NS

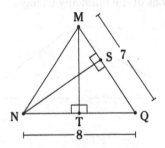

Balancing Things Out with Medians and Centroids

Because triangles aren't as symmetrical as, say, circles or rectangles (except for the equilateral triangle), they don't have an obvious center point like circles and rectangles do. In this and the next section, you look at four different "centers" that every triangle has. Of the four, the centroid is probably the best candidate for a triangle's true center.

REMEMBER

A triangle's medians point the way to its centroid. Here's everything you've always wanted to know about medians but were afraid to ask.

» **Median:** A *median* of a triangle is a segment joining a *vertex* (corner point) with the midpoint of the opposite side. Every triangle has three medians.

» **Centroid:** The three medians of a triangle intersect at a single point called the *centroid*. (The centroid is the triangle's center of gravity, or balance point.)

» **Position of centroid on median:** Along every median, the distance from the vertex to the centroid is twice as long as the distance from the centroid to the midpoint.

EXAMPLE

Q. Given $\triangle BSF$ with medians $\overline{BH}$, $\overline{SU}$, and $\overline{FA}$ and centroid L

a. If FL is 12, what's FA?

b. If BH is 12, what's HL?

c. If SL is 12, what's UL?

d. If the area of $\triangle BSF$ is 20 units^2, what's the area of $\triangle BSU$?

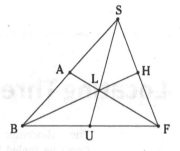

A. The centroid, L, cuts each median into a $\frac{1}{3}$ part and a $\frac{2}{3}$ part. Notice that it's obvious from the figure which is the short part and which is the long part. (Try measuring the parts with your fingers.)

a. FL is $\frac{2}{3}$ of FA, so if FL is 12, $FA = 18$.

b. HL is $\frac{1}{3}$ of BH, so if BH is 12, $HL = 4$.

c. A centroid is twice as far from a vertex as it is from the midpoint of the opposite side, so $\overline{SL}$ is twice as long as $\overline{UL}$. $SL = 12$, so $UL = 6$.

d. $\triangle BSF$ and $\triangle BSU$ have the same altitude (it goes from point S straight down to $\overline{BF}$, hitting $\overline{BF}$ somewhere between U and F). $\overline{SU}$ is a median, so U is the midpoint of $\overline{BF}$. Thus, $\overline{BU}$, the base of $\triangle BSU$, is half as long as $\overline{BF}$, the base of $\triangle BSF$. Therefore, because their altitudes are the same, and because $\triangle BSU$ has a base that's half of the base of $\triangle BSF$, the area of $\triangle BSU$ must be half of the area of $\triangle BSF$. The answer is 10 units^2.

10 Draw in the medians of △*ABC*. Do they appear to bisect the vertex angles?

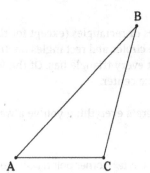

 *11 $\overline{NT}$ and $\overline{HO}$ are medians of △*NRH*. If the area of △*NOH* is 13, what's the area of △*NRT*?

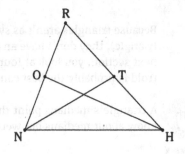

Locating Three More "Centers" of a Triangle

The orthocenter, incenter, and circumcenter are three points associated with every triangle. Don't be fooled by the term "center," though. You can see in a minute why they're called centers, but it's not because these points are near the center of the triangle.

REMEMBER

Here's a brief description of each "center."

>> **Orthocenter:** Where a triangle's three *altitudes* intersect.

>> **Incenter:** Where a triangle's three *angle bisectors* intersect; it's the center of a circle *inscribed* in (drawn inside) the triangle.

>> **Circumcenter:** Where the three *perpendicular bisectors* of the sides intersect; it's the center of a circle *circumscribed* about (drawn around) the triangle.

Well, I guess you only get to see why the incenter and the circumcenter are called centers. I don't know why the orthocenter is called a center, but two out of three ain't bad.

If you sketch a few differently shaped triangles, you can see that there isn't always an obvious place where you'd say the center is, like there would be with, for example, a rectangle. I think a triangle's centroid is the best choice for a triangle's center — better than the so-called "centers" just mentioned. Here's why: The centroid is the triangle's center of gravity, and it always seems to be near what common sense would say is the center.

Location of the centroid and the three "centers." Of the three "centers" described in this section, two of them (the orthocenter and circumcenter) are sometimes outside of the triangle. The third one (the incenter) is sometimes way at one end of the triangle. Here's the lowdown for the three "centers" plus the centroid:

REMEMBER

>> For all types of triangles, the centroid and incenter are inside the triangle.

>> In an acute triangle, the orthocenter and circumcenter are inside the triangle as well.

>> In a right triangle, the orthocenter and circumcenter are on the triangle.

>> In an obtuse triangle, the orthocenter and circumcenter are outside the triangle.

Here's a mnemonic device to help you keep the four "centers" straight. It's admittedly not one of my better mnemonics, but it'll probably work just fine, and it's certainly better than nothing. First, pair up the four "centers" with the lines, rays, or segments that intersect:

TIP

>> Centroid — Medians

>> Circumcenter — Perpendicular bisectors

>> Incenter — Angle bisectors

>> Orthocenter — Altitudes

Notice that the two terms on the left that begin with consonants pair up with terms on the right that begin with consonants. Ditto for the terms that begin with vowels. The only two terms that contain double vowels (*oi* and *ia*) are paired up. And the two terms with two *t*'s (*orthocenter* and *altitude*) go together. Easy, right?

Q. In the following triangles, identify all marked centroids, orthocenters, incenters, and circumcenters. Try this exercise on your own before reading the solution.

a)

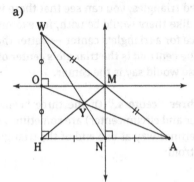

b)

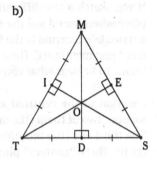

c)

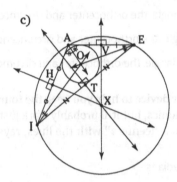

A. In △HWA, the tick marks tell you that M, N, and O are midpoints; therefore, $\overline{HM}$, $\overline{WN}$, and $\overline{AO}$ are medians. Y is thus the centroid. The right-angle marks at the midpoints tell you that $\overline{OM}$ and $\overline{NM}$ are perpendicular bisectors of sides $\overline{HW}$ and $\overline{HA}$. They cross at M, so M is the triangle's circumcenter. (Note that you don't need the third perpendicular bisector; you know that all three intersect at the same point, so any two can show you where the circumcenter is.) Finally, (this one's a bit tricky), you identify point H as the orthocenter. Missing this point is easy because H is part of the triangle. But you can see that $\overline{HW}$ and $\overline{HA}$ are altitudes of △HWA (the two legs of a right triangle are always altitudes), and because $\overline{HW}$ and $\overline{HA}$ intersect at H, H has to be the orthocenter. The incenter of △HWA does not appear on this figure.

Points E, D, and I in △TMS are marked as midpoints, and thus, $\overline{TE}$, $\overline{MD}$, and $\overline{SI}$ are medians. They cross at O, so O is the centroid. The right angle marks on the figure and the tick marks on the angles tell you that $\overline{TE}$, $\overline{MD}$, and $\overline{SI}$ are also altitudes *and* angle bisectors *and* perpendicular bisectors. So, yup, point O is all four points wrapped up into one: the centroid, the orthocenter, the incenter, and the circumcenter. By the way, this overlap happens only in an equilateral triangle. In fact, the four points are *always* four distinct points except when they all come together in an equilateral triangle.

For △IAE, the two circles should make this a no-brainer. Point O, the center of the inscribed circle, is, by definition, the incenter. And point X, the center of the circumscribed circle, is, by definition, the circumcenter. Neither the orthocenter nor the centroid of △IAE appears on this figure.

12 Pick and choose: Identify the centroid, the orthocenter, the incenter, and the circumcenter in △*XYZ*. This figure is drawn to scale.

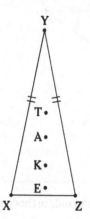

13 Pick and choose: Identify the centroid, the orthocenter, the incenter, and the circumcenter in △*ABC*. This figure is drawn to scale.

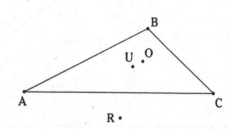

14 Pick and choose: Identify the centroid, the orthocenter, the incenter, and the circumcenter in △*STU*. This figure is drawn to scale.

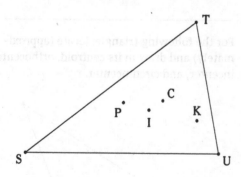

15 What does the fact that *Y* and *R* are outside the triangle tell you (this is the same figure as in problem 13)?

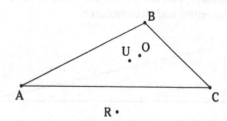

16 For the following triangle, locate (approximately) and draw in its centroid, orthocenter, incenter, and circumcenter. *Hint for problems 16 to 19*: Just sketch two medians to find the centroid, two perpendicular bisectors to find the circumcenter, and so on.

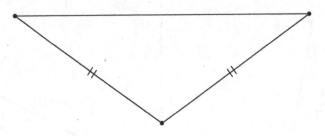

17 For the following triangle, locate (approximately) and draw in its centroid, orthocenter, incenter, and circumcenter.

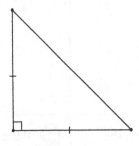

18 For the following triangle, locate (approximately) and draw in its centroid, orthocenter, incenter, and circumcenter.

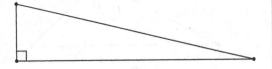

19 For the following triangle, locate (approximately) and draw in its centroid, orthocenter, incenter, and circumcenter.

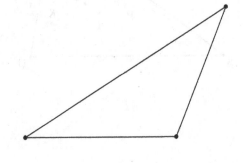

The Pythagorean Theorem

Drum roll, please. Ladieeeees and gentlemen, in the center ring, for your enjoyment and amazement, all the way from Samos, Greece, from over 2600 years ago, I bring you . . . the Pythagorean Theorem! Pretty thrilling, eh?

The Pythagorean Theorem is certainly one of the most famous theorems in all of mathematics. Mathematicians and lay people alike have studied it for centuries. People have proved it in many different ways. Even President James Garfield was credited with a new, original proof. Well, here you go. As the Scarecrow in *The Wizard of Oz* tried to say after he got his Doctor of Thinkology diploma (a "Th.D.") to prove he had brains. . .

THEOREMS & POSTULATES

The Pythagorean Theorem: The sum of the squares of the legs of a right triangle is equal to the square of the hypotenuse.

(Actually, the Scarecrow misstated it as, "The sum of the square roots of any two sides of an isosceles triangle is equal to the square root of the remaining side.")

Figure 4-2 contains the well-known $3 - 4 - 5$ triangle to visually show you the meaning of the Pythagorean Theorem.

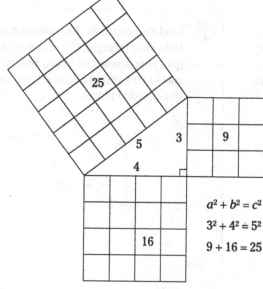

FIGURE 4-2:
Nine little squares plus 16 little squares equals 25 little squares. Pythagoras, Pyschmagoras— what's all the fuss about?

$$a^2 + b^2 = c^2$$
$$3^2 + 4^2 = 5^2$$
$$9 + 16 = 25$$

Q. Calculate the length of the unknown sides in the triangles to the right.

EXAMPLE

a)

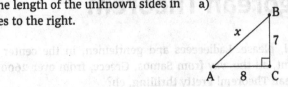

b)

A. $\triangle ABC: a^2 + b^2 = c^2$

$$7^2 + 8^2 = x^2$$
$$49 + 64 = x^2$$
$$113 = x^2$$
$$x = \sqrt{113}$$
$$\approx 10.6$$

$\triangle XYZ: a^2 + b^2 = c^2$

$$12^2 + y^2 = 13^2$$
$$144 + y^2 = 169$$
$$y^2 = 25$$
$$y = 5$$

---

20 Find the length of the unknown side in the following triangle. If the answer is irrational, give your answer in exact, radical (square root) form and in decimal form rounded to two decimal places.

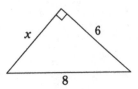

21 Find the length of the unknown side in the following triangle. If the answer is irrational, give your answer in exact, radical (square root) form and in decimal form rounded to two decimal places.

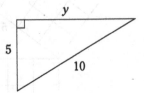

22 Find the length of the unknown side in the following triangle. If the answer is irrational, give your answer in exact, radical (square root) form and in decimal form rounded to two decimal places.

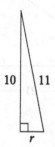

23 Find the length of the unknown side in the following triangle. If the answer is irrational, give your answer in exact, radical (square root) form and in decimal form rounded to two decimal places.

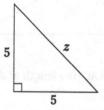

24 Find the length of the unknown side in the following triangle. If the answer is irrational, give your answer in exact, radical (square root) form and in decimal form rounded to two decimal places.

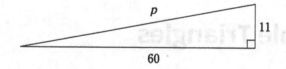

25 Find the length of the unknown side in the following triangle. If the answer is irrational, give your answer in exact, radical (square root) form and in decimal form rounded to two decimal places.

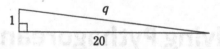

26 Find x.

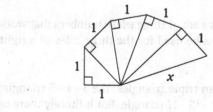

27 Find PS, SR, PR, and the area of $\triangle PQR$.

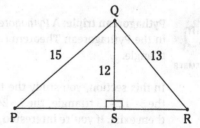

28 Answer the following questions using this figure:

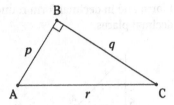

a. Express AC (the length of $\overline{AC}$) in terms of p and q.

b. Express AB in terms of q and r.

c. Express BC in terms of p and r.

*29 Find the area of △MOJ without using Hero's formula.

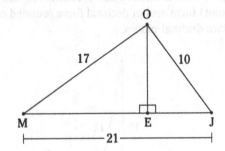

Solving Pythagorean Triple Triangles

If you pick any old numbers for two of the sides of a right triangle, the third side usually ends up being irrational — you know, the square root of something. For example, if the legs are 5 and 8, the hypotenuse ends up being $\sqrt{5^2 + 8^2} = \sqrt{89} \approx 9.43398...$ (the decimal goes on forever without repeating). And if you pick whole numbers for the hypotenuse and one of the legs, the other leg usually winds up being the square root of something.

When this doesn't happen — namely, when all three sides are whole numbers — you've got a Pythagorean triple.

REMEMBER

Pythagorean triple: A *Pythagorean triple* (like $3 - 4 - 5$) is a set of three *whole* numbers that work in the Pythagorean Theorem ($a^2 + b^2 = c^2$) and can thus be used for the three sides of a right triangle.

In this section, you study the four smallest Pythagorean triple triangles: the $3 - 4 - 5$ triangle; the $5 - 12 - 13$ triangle; the $7 - 24 - 25$ triangle; and the $8 - 15 - 17$ triangle. But infinitely more of them exist. If you're interested, one simple way to find more of them is to take any odd number, say 11, and square it — that's 121. The two consecutive numbers that add up to 121 (60 and 61) give you the two other numbers (to go with the 11). So, another Pythagorean triple is $11 - 60 - 61$.

A *family* of right triangles is associated with each Pythagorean triple. For example, the $5:12:13$ family consists of the $5-12-13$ triangle and all other triangles of the same shape that you'd get by shrinking or blowing up the $5-12-13$ triangle. If you shrink it 100 times, you get a $5/100-12/100-13/100$ triangle. Or you can quadruple each side and get a $20-48-52$ triangle or multiply each side by $\sqrt{17}$ to get a $5\sqrt{17}-12\sqrt{17}-13\sqrt{17}$ triangle.

Q. Find the lengths of the unknown sides in the following triangles by looking for triangle families. (Don't use the Pythagorean Theorem.)

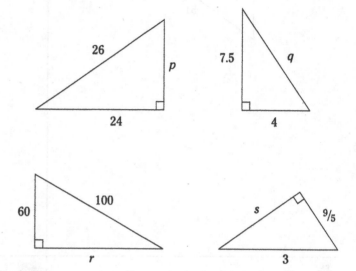

A. For the *p* triangle, you want to first notice that 26 is twice 13. That should ring the $5:12:13$ bell. Then you check that 24 is twice 12, which, of course, it is. Thus, you have a $5-12-13$ triangle blown up to twice its size; therefore, *p* is $2 \cdot 5$, or 10.

For the *q* triangle, you recognize the triangle family if you get rid of that pesky decimal. You can do that by multiplying the 7.5 and the 4 by 2, which gives you 15 and 8. Bingo — you have an $8-15-17$ triangle shrunk in half. So, *q* is half of 17, or 8.5.

For the *r* triangle, first divide the 60 and 100 by 10 — that's 6 and 10. This should ring the $3:4:5$ bell. Doubling 3, 4, and 5 gives you 6, 8, and 10, and then multiplying by 10 gives you 60, 80, and 100, so *r* is 80.

Finally, for the *s* triangle, multiply the 3 and the 9/5 by the denominator, 5, to get 15 and 9. Then reduce these terms by dividing each by 3: That gives you 5 and 3, and, voilà, you have a triangle in the $3:4:5$ family. One neat way to find *s* is to now take the 4 (because the two given sides became the 5 and 3) and reverse the process: *multiply* by 3 (that's 12) and then *divide* by 5: *s* is 12/5.

30 Without using the Pythagorean Theorem, find the length of the unknown side in the following triangle.

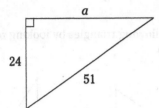

31 Without using the Pythagorean Theorem, find the length of the unknown side in the following triangle.

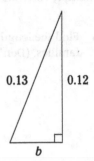

32 Without using the Pythagorean Theorem, find the length of the unknown side in the following triangle.

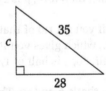

33 Without using the Pythagorean Theorem, find the length of the unknown side in the following triangle.

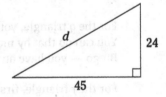

Without using the Pythagorean Theorem, find the length of the unknown side in the following triangle.

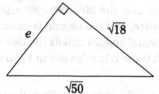

Without using the Pythagorean Theorem, find the length of the unknown side in the following triangle.

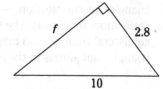

*36 **Find a, b, c, and d.**

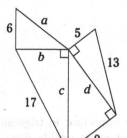

37 **Find x.**

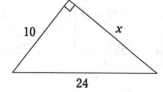

38 **Find c.**

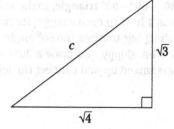

Unique Degrees: Two Special Right Triangles

REMEMBER

The Pythagorean triple families of triangles you find in the last section are nice to know because they come up in so many right triangle problems. But mathematically speaking, the two right triangles in this section — the 45°–45°–90° triangle and the 30°–60°–90° triangle — are really more important. The first is exactly half of a square, and the second is exactly half of an equilateral triangle, and this connection to those elemental shapes makes the two special right triangles ubiquitous in the geometry landscape. Check these triangles out in Figure 4-3:

» **The 45°–45°–90° triangle** has angles of 45°, 45°, and 90° (duh) and sides in the ratio of $1:1:\sqrt{2}$. This triangle is the shape of half a square, cut along its diagonal.

» **The 30°–60°–90° triangle** has angles of 30°, 60°, and 90° and sides in the ratio of $1:\sqrt{3}:2$. This triangle is the shape of half an equilateral triangle cut down the middle along its altitude.

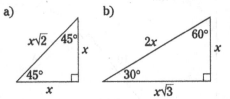

FIGURE 4-3:
Two special right triangles.

The 45°–45°–90° and 30°–60°–90° triangles are very important in trigonometry and to a lesser extent in calculus. Get to know them forwards, backwards, upside-down, and sideways.

WARNING

Don't mix up the 2x and the $x\sqrt{3}$. When you use the Pythagorean Theorem, you often end up with a hypotenuse with a square root in it. Because of this, students often mix up the 2x and the $x\sqrt{3}$ for the 30°–60°–90° triangle and put the $x\sqrt{3}$ on the hypotenuse. You can avoid this mistake if you remember that $\sqrt{3}$ is less than 2 (think for a few seconds and figure out why it has to be less than 2); because the hypotenuse is always the longest side of a right triangle, the 2x has to go on the hypotenuse.

TIP

Whenever you sketch a 30°–60°–90° triangle, make sure you make the long leg much longer than the short leg (it doesn't hurt to even exaggerate the relationship a bit). That way, it'll be obvious to you that the short leg touches the 60° angle and that the long leg touches the 30° angle. If you instead get a bit sloppy and draw a 30°–60°–90° triangle so that the legs look about equal, it's easy to get mixed up and connect the legs to the wrong angles.

EXAMPLE

Q. Find the lengths of the unknown sides in $\triangle CBA$ and $\triangle WQX$.

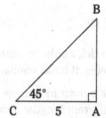
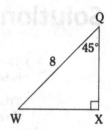

A. You have two ways to solve these problems that really amount to the same thing. First, you can use the ratio of the sides of the $45° - 45° - 90°$ triangle from Figure 4-3:

leg : leg : hypotenuse

x : x : $x\sqrt{2}$

In $\triangle CBA$, one of the legs is 5, so x is 5. Now just plug 5 into $x : x : x\sqrt{2}$ and you have the three sides: 5, 5, and $5\sqrt{2}$.

In $\triangle WQX$, the hypotenuse is 8, so you set $x\sqrt{2}$ equal to 8 and solve for x:

$$x\sqrt{2} = 8$$
$$x = \frac{8}{\sqrt{2}} = \frac{8\sqrt{2}}{2} = 4\sqrt{2}$$

So, the three sides are $4\sqrt{2}$, $4\sqrt{2}$, and 8.

I prefer the following method: Just think of the $45° - 45° - 90°$ triangle as the $\sqrt{2}$ triangle (or "root 2 triangle"). Now, if you know the length of a leg and you want the length of the hypotenuse (a *longer* thing), you *multiply* by $\sqrt{2}$. And if you know the hypotenuse and want to figure a leg (a *shorter* thing), you *divide* by $\sqrt{2}$. That's all there is to it.

 Find the area of an equilateral triangle whose sides are 10.

 Find the area of a square whose diagonal has a length of 10.

Solutions

1 The two tick marks in triangle a tell you that those two sides are equal and, thus, the triangle is isosceles. It looks equilateral, but you can't assume that.

Triangle b must be scalene, because no matter what x is, x and $x + 1$ and $x + 2$ will always be three different lengths. Don't be fooled by the fact that the triangle looks isosceles.

2 Because $\triangle LSO$ is isosceles, at least two of the sides must be equal. First try $\overline{IS} \cong \overline{SO}$:

$$2x = 4x - 4$$
$$-2x = -4$$
$$x = 2$$

Plugging $x = 2$ into the three sides gives you sides of 4, 4, and 1; that's not a large enough total because the perimeter is supposed to be more than 10. Try $\overline{SO} \cong \overline{IO}$:

$$4x - 4 = 3x - 5$$
$$x = -1$$

No good. This setup gives you three sides of negative length.

The third pair better work, because that's the only thing left to try:

$$\overline{IO} \cong \overline{IS}$$
$$3x - 5 = 2x$$
$$x = 5$$

Plugging $x = 5$ into the three sides gives you sides of length 10, 10, and 16. Bingo. The base, $\overline{SO}$, is 16 and the legs, $\overline{IO}$ and $\overline{IS}$, are both 10.

3 The angles are in the ratio of $4 : 5 : 6$, so set the angles equal to $4x$, $5x$, and $6x$. The angles in a triangle add up to $180°$, so

$$4x + 5x + 6x = 180$$
$$15x = 180$$
$$x = 12$$

Plugging $x = 12$ into $4x$, $5x$, and $6x$ gives you three acute angles, $48°$, $60°$, $72°$, so it's an acute triangle. And because the triangle has three unequal angles, it must have three unequal sides as well. So, it's scalene.

4 For the first triangle, the supplement of the $140°$ angle is $40°$, and the vertical angle across from the $50°$ angle is, of course, also $50°$. So far, you have a $40°$ angle and a $50°$ angle. The third angle has to give you a total of $180°$, so the third angle is $90°$: You have a right triangle.

Did you think the second triangle was obtuse? Good try, but look again. This isn't any type of triangle — not in our universe anyway — because the angles don't add up to $180°$.

5 Here are the answers.

a. Always: An equilateral triangle is isosceles by definition.

b. Sometimes: An isosceles triangle is equilateral only when its base is congruent to its legs.

c. Sometimes: A right triangle is isosceles when its legs are congruent (in other words, when it's a $45° - 45° - 90°$ triangle — see the section, "Unique Degrees: Two Special Right Triangles," in this chapter).

d. Always: $70°$ plus $55°$ is $125°$. The third angle must bring the total to $180°$, so it's another $55°$ angle, and therefore, the triangle is isosceles.

e. Sometimes: If the vertex angle of an obtuse isosceles triangle is $100°$, its base angles will both be $40°$, so the answer has to be at least *sometimes*. But the answer isn't *always*, because the vertex angle of an obtuse isosceles triangle can have any measure greater than $90°$ and less than $180°$.

f. Never: $40°$ plus $40°$ is $80°$, so the third angle must be $100°$, which makes the triangle obtuse.

g. Never: A triangle can never have two supplementary angles, because they would add up to $180°$ and there'd be nothing left for the third angle.

h. Never: If two of the angles in a triangle are complementary, they add up to $90°$, and that leaves $90°$ for the third angle, because all three angles have to total $180°$. Thus, the triangle must be a right triangle.

6 Your answer should look roughly like this:

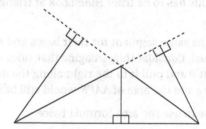

You may want to spin this figure around to make the dotted lines horizontal (like a tabletop) and the altitudes going straight up from the table. That's a good way to picture altitudes and to see where they should go.

7 The area equals 30 square units:

$$\text{Area}_\triangle = \frac{1}{2}\, \text{base} \cdot \text{height}$$
$$= \frac{1}{2} \cdot 13 \cdot \frac{60}{13}$$
$$= 30$$

Now, if you instead use the 12 as the base, the altitude is 5:

$$\text{Area}_\triangle = \frac{1}{2}\, 12 \cdot 5$$
$$= 30$$

If 5 is the base, the height is 12:

$$\text{Area}_\triangle = \frac{1}{2} \cdot 5 \cdot 12$$
$$= 30$$

Finally, use Hero's formula:

$$S = \frac{5 + 12 + 13}{2} = 15$$
$$\text{Area}_\triangle = \sqrt{S(S-a)(S-b)(S-c)}$$
$$= \sqrt{15(15-5)(15-12)(15-13)}$$
$$= \sqrt{15(10)(3)(2)}$$
$$= \sqrt{900}$$
$$= 30$$

(8) The area of rectangle $ABDE$ is, of course, $6 \cdot 10$, or 60 units^2. The area of $\triangle ACE$ is $\frac{1}{2} \cdot 6 \cdot 10$, or 30 units^2. $\triangle AGE$ and $\triangle APE$ both have the same base as $\triangle ACE$ (namely $\overline{AE}$), and, like $\triangle ACE$, they both have a height of 10 (the vertical distance between the two dotted lines). Thus, the areas of the three triangles are the same.

You can draw these two conclusions:

- The area of a triangle is half of the area of a rectangle with that same base and height. (Do you see why this has to be true? *Hint:* Look at triangles ACF and CAB and triangles ECF and EDC.)

- If triangles have the same segment for their bases and their heights are the same, then their areas are equal. Consider this: Imagine that sides $\overline{AP}$ and $\overline{EP}$ are made of elastic, and you grab point P and pull it to the right along the dotted line. You could pull it out 1000 miles or more and the area of $\triangle APE$ would still be only 30 units^2.

(9) For this one, you have to use the area formula twice:

$$\text{Area}_\triangle = \frac{1}{2} b \cdot h$$
$$= \frac{1}{2} \cdot 8 \cdot 6$$
$$= 24$$

Now, because $\overline{NS}$ is the altitude drawn to base $\overline{MQ}$, you can figure NS by using the area formula backwards:

$$\text{Area}_\triangle = \frac{1}{2} \cdot b \cdot h$$
$$24 = \frac{1}{2} \cdot 7 \cdot h$$
$$48 = 7 \cdot h$$
$$h = \frac{48}{7}$$

So, $NS = \frac{48}{7}$.

10 Here's what your figure should look like:

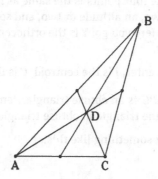

Be careful with this one. Many people take a quick look at the medians and say that it *does* look like they bisect the vertex angles. I hope you didn't jump to that conclusion. If you look carefully, you can see that although ∠ABC looks like it might be cut in half, the medians from A and C don't even come close to bisecting their angles. (∠ACD in the figure looks like it's somewhere around a 70° angle. But ∠DCB, on the other hand, looks more like a 45° angle.)

It turns out that ∠B isn't bisected either, though it's pretty close. Only the median to the base of an isosceles triangle bisects the vertex angle (and therefore, all three medians of an equilateral triangle bisect the vertex angles).

*11 At first, you may feel that you've got nothing to go on to solve this problem. You may be thinking, "How can I get the area of △NRT when I don't know anything about it?"

Well, the logic here is quite similar to the reasoning in part *d* of the example problem. Because $\overline{HO}$ is a median, O is a midpoint, and thus $\overline{NR}$ is twice as long as $\overline{HO}$. Now spin the triangle so that $\overline{NR}$ becomes the base. You can see that △NRH and △NOH have the same height. Because the base of △NRH is twice as long as the base of △NOH, the area of △NRH is twice the area of △NOH — so the area of △NRH is 26 units^2.

With the same reasoning — this time spinning the triangle so that $\overline{RH}$ is the base — you can conclude that the area of △NRT is half the area of the whole triangle. So, like △NOH, the area of △NRT is 13 units^2.

12 In △XYZ, if you very roughly sketch the perpendicular bisector of $\overline{XY}$ or $\overline{ZY}$, you can see that it crosses T and doesn't even come close to A, K, or E. So, T has to be the circumcenter.

If you sketch the median from Y straight down the middle of the triangle, it passes through all four points. Recalling that the centroid is at the 1/3 point of each median, you can see that point A has to be the centroid. (K and E are nowhere near 1/3 of the way up the median.)

Next, sketch an altitude from, say, angle X perpendicular to $\overline{YZ}$. This segment passes through point E (or close to E, depending on how good your sketching skills are), so E has to be the orthocenter. And if that doesn't convince you, K can't possibly be the orthocenter, because if you draw a line through angle X and point K and that crosses over $\overline{YZ}$, it's easy to see that they're not perpendicular and that, therefore, K is not on an altitude.

Now that you've found the first three points, you have no other choice for K — it has to be the incenter.

(13) Here, I skip the lengthy explanation for △*ABC* (and △*STU* in the next problem) because the method for identifying the four points is the same as for △*XYZ* in the preceding problem. You just sketch a median or two, an altitude or two, and so on until, by process of elimination, you've made your picks. Here you go: *Y* is the orthocenter, *O* is the incenter, *U* is the centroid, and *R* is the circumcenter.

(14) Pick 'em: *P* is the circumcenter, *I* is the centroid, *C* is the incenter, and *K* is the orthocenter.

(15) You can conclude that △*ABC* is an obtuse triangle. Remember that the orthocenter and the circumcenter are *outside* the triangle in obtuse triangles.

(16) Your solution should look something like this:

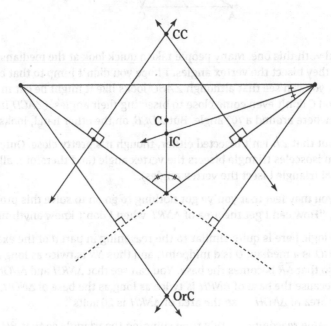

(17) Your solution should look something like this:

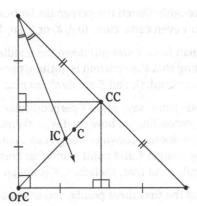

18 Your solution should look something like this:

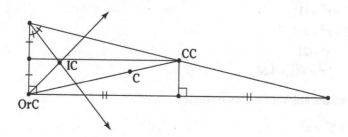

19 Your solution should look something like this:

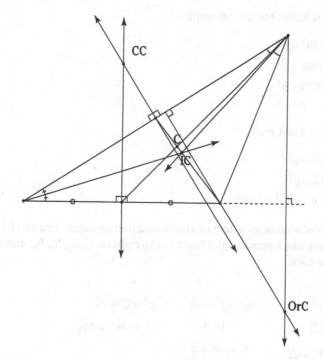

20 You use, of course, $a^2 + b^2 = c^2$ for all six triangles in problems 20 to 25.

$$x^2 + 6^2 = 8^2$$
$$x^2 + 36 = 64$$
$$x^2 = 28$$
$$x = \sqrt{28} = 2\sqrt{7} \approx 5.298$$

21 Here you go:

$$y^2 + 5^2 = 10^2$$
$$y^2 = 75$$
$$y = \sqrt{75} = 5\sqrt{3} \approx 8.66$$

(22) Take a look at the answer for the r triangle:

$$10^2 + r^2 = 11^2$$
$$100 + r^2 = 121$$
$$r^2 = 21$$
$$r = \sqrt{21} \approx 4.58$$

(23) Check out the answer for this triangle:

$$5^2 + 5^2 = z^2$$
$$50 = z^2$$
$$z = \sqrt{50} = 5\sqrt{2} \approx 7.07$$

(24) Here's the solution for the p triangle:

$$11^2 + 60^2 = p^2$$
$$121 + 3600 = p^2$$
$$3721 = p^2$$
$$p = 61$$

(25) And here's the last one:

$$1^2 + 20^2 = q^2$$
$$1 + 400 = q^2$$
$$q = \sqrt{401} \approx 20.02$$

(26) This is sort of a domino-effect or chain-reaction problem. You can label the hypotenuses (or is it *hypoteni*, like *hippopotami*?) from left to right as h_1, h_2, h_3, h_4, and x. Now the problem's a walk in the park:

$$h_1^2 = 1^2 + 1^2$$
$$h_1 = \sqrt{2}$$
$$h_2^2 = 1^2 + \sqrt{2}^2$$
$$= 1 + 2$$
$$h_2 = \sqrt{3}$$

$$h_3^2 = 1^2 + \sqrt{3}^2$$
$$= 1 + 3$$
$$h_3 = \sqrt{4} = 2$$
$$h_4^2 = 1^2 + 2^2$$
$$= 1 + 4$$
$$h_4 = \sqrt{5}$$

$$x^2 = 1^2 + \sqrt{5}^2$$
$$x = \sqrt{6} \approx 2.45$$

(27) Here are your answers:

$$PS^2 + 12^2 = 15^2 \qquad SR^2 + 12^2 = 13^2 \qquad PR = PS + SR$$
$$PS^2 + 144 = 225 \qquad SR^2 + 144 = 169 \qquad = 9 + 5$$
$$PS^2 = 81 \qquad\qquad SR^2 = 25 \qquad\quad PR = 14$$
$$PS = 9 \qquad\qquad\quad SR = 5$$

$$\text{Area}_{\Delta PQR} = \frac{1}{2}\,\text{base}\cdot\text{height}$$
$$= \frac{1}{2}\cdot 14 \cdot 12$$
$$= 84 \text{ units}^2$$

(28) You know $a^2 + b^2 = c^2$; therefore,

a. $AC^2 = p^2 + q^2$
$$AC = \sqrt{p^2 + q^2}$$

b. $AB^2 + q^2 = r^2$
$$AB^2 = r^2 - q^2$$
$$AB = \sqrt{r^2 - q^2}$$

c. $p^2 + BC^2 = r^2$
$$BC^2 = r^2 - p^2$$
$$BC = \sqrt{r^2 - p^2}$$

(*29) Label the altitude h and let $\overline{EJ}$ equal x; $\overline{ME}$ is $21 - x$. Then use the Pythagorean Theorem for both triangles and solve the system of two equations with two unknowns.

$$\Delta MOE: \qquad h^2 + (21 - x)^2 = 17^2$$
$$h^2 + 441 - 42x + x^2 = 289$$
$$h^2 + x^2 - 42x = -152$$

$$\Delta JOE: \quad h^2 + x^2 = 10^2$$
$$h^2 + x^2 = 100$$

Now subtract the second equation from the first:

$$h^2 + x^2 - 42x = -152$$
$$\underline{-\left(h^2 + x^2 = 100\right)}$$
$$-42x = -252$$
$$x = \frac{-252}{-42}$$
$$x = 6$$

Next, plug $x = 6$ into the equation for $\triangle JOE$ to get h:

$$h^2 + 6^2 = 100$$
$$h^2 = 64$$
$$h = 8$$

Finally, finish with the area formula:

$$\text{Area}_{\triangle MOJ} = \frac{1}{2}bh$$
$$= \frac{1}{2} \cdot 21 \cdot 8$$
$$= 84 \text{ units}^2$$

(30) For the a triangle, reduce the $24:51$ ratio — you know, just like reducing a fraction. Dividing each number by 3, the ratio reduces to $8:17$, so you have a triangle in the $8:15:17$ family. The a is the missing 15 side, and because the triangle is an $8-15-17$ triangle blown up 3 times, a is $3 \cdot 15$, or 45.

(31) The b triangle should be a no-brainer, because you have a 12 (namely 0.12) and a 13 (namely 0.13) staring you in the face. You know b is thus 0.05, and the triangle is, of course, a $5-12-13$ triangle shrunk down 100 times.

(32) Do the c triangle just like the a triangle in problem 30. Reducing $28:35$ by 7 gives you $4:5$. Bingo: It's a $3:4:5$ triangle. You find that c is the missing 3 side. Blowing the side back up 7 times gives you 21 for c.

(33) The d triangle is also just like the a triangle from problem 30. For the d triangle, the greatest common factor of 24 and 45 is 3. Reducing by 3 gives you 8 and 15, so the triangle's in the $8:15:17$ family. You determine that d is 17 times 3, or 51.

(*34) I give you two ways to solve the e triangle. The first method involves simplifying the radicals:

$$\sqrt{18} = \sqrt{9 \cdot 2} = \sqrt{9}\sqrt{2} = 3\sqrt{2} \quad \text{and}$$
$$\sqrt{50} = \sqrt{25 \cdot 2} = \sqrt{25}\sqrt{2} = 5\sqrt{2}$$

The $3\sqrt{2}$ and $5\sqrt{2}$ tell you that you have a $3-4-5$ triangle blown up $\sqrt{2}$ times. The missing side, e, is thus $4\sqrt{2}$.

The second method is to take your calculator, enter $\frac{\sqrt{18}}{\sqrt{50}}$, and hit *Enter* or =. You get an answer of 0.6. Then "fraction" that, and you get 3/5: Voilà, your triangle is a $3:4:5$ triangle. Now enter $\frac{\sqrt{50}}{5}$ to find the blow-up multiplier — it's about 1.41. Your approximate answer is 4 times 1.41, or 5.64. (For problems with square roots, this second method can give you only an approximate answer unless you have a super-duper calculator, like the TI-Nspire.)

(*35) Just use the calculator trick for the f triangle. Enter $2.8 \div 10$ and hit *Enter*. That gives you 0.28. "Fraction" that and you get 7/25. Bingo. It's a $7:24:25$ triangle. Dividing 25 by 10 gives you a shrink factor of 2.5. Finally, f equals 24 divided by 2.5, which is 9.6.

(*36) I suspect you figured out that you've got to solve for d first, then c, and so on. You know d is, of course, 12. Then that 12 and the 9 are two legs of a $3-4-5$ triangle blown up 3 times.

Thus, c is 15; b is then 8, of course. Finally, that 8 and the 6 are two legs of another triangle in the $3:4:5$ family, so a is 10.

(37) The 10 and the 24 are a 5 and a 12 doubled, so that should ring a bell — the $5:12:13$ bell. But don't answer the bell! In a $5:12:13$ triangle, the 13 represents the hypotenuse, but in this triangle, the 24 (which corresponds to the 12) is the hypotenuse. So, this triangle doesn't belong to any of the Pythagorean triple families. You have to solve this triangle with the Pythagorean Theorem:

$$10^2 + x^2 = 24^2$$
$$100 + x^2 = 576$$
$$x^2 = 476$$
$$x = \sqrt{476} \approx 21.82$$

(38) Another tricky question. Did you conclude that c equals $\sqrt{5}$? This is *not* a $3:4:5$ triangle, which you can check on your calculator. The ratio of the two legs in a $3:4:5$ triangle is $3/4$, or 0.75. But if you do $\frac{\sqrt{3}}{\sqrt{4}}$ on your calculator, you get something different (0.866), which shows that this triangle is not in the $3:4:5$ family. Solve with the Pythagorean Theorem:

$$c^2 = \sqrt{3}^2 + \sqrt{4}^2$$
$$c^2 = 3 + 4$$
$$c^2 = 7$$
$$c = \sqrt{7}$$

(39) Draw your equilateral triangle with its altitude like this:

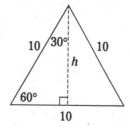

You have the base, so all you need to compute the area is the height. Half of an equilateral triangle is a $30° - 60° - 90°$ triangle, and you can see that h is the long leg. The short leg is 5; multiply that by $\sqrt{3}$ to get h; h is $5\sqrt{3}$. Finish with the area formula:

$$\text{Area} = \frac{1}{2}bh$$
$$= \frac{1}{2} \cdot 10 \cdot 5\sqrt{3}$$
$$= 25\sqrt{3} \text{ units}^2$$

You can also, of course, solve this problem with the formula for the area of an equilateral triangle from the section about altitudes and area ($A = \frac{s^2\sqrt{3}}{4}$). But it's not a bad idea to know the preceding method using the $30° - 60° - 90°$ triangle because it's useful in its own right. Plus, this method can really come in handy in case you forget the formula.

(*40) Half a square cut along its diagonal is a $45° - 45° - 90°$ triangle. The square's diagonal is the hypotenuse of the $45° - 45° - 90°$ triangle. That's 10, so you divide 10 by $\sqrt{2}$ to get the sides of the square — $\frac{10}{\sqrt{2}}$. The area of a square is, of course, s^2, so this square has an area of $\left(\frac{10}{\sqrt{2}}\right)^2 = \frac{100}{2} = 50 \text{ units}^2$.

Chapter 5

Proofs Involving Congruent Triangles

In this chapter, you dive into proofs in a big way. The triangle proofs you do in this and subsequent chapters are real, full-fledged proofs. The proofs in Chapter 3 are basically just warm-up exercises for the longer proofs you do from here on. However, I don't want to diminish the importance of the Chapter 3 material. In fact, Chapter 3 is where you practice using the important theorems and proof techniques that you need for longer proofs, so it's critical that you understand that material before you continue. If you understand Chapter 3, the proofs in this and later chapters probably won't cause you too much trouble (perhaps just the occasional brain hemorrhage).

Sizing Up Three Ways to Prove Triangles Congruent

In a proof, the point at which you prove triangles congruent is sort of like the climax in a novel: Everything builds up to it, and it's the focus or main point or anchor of the proof. Some of the shorter proofs in this chapter end with proving triangles congruent. In longer, more typical proofs, you take things to the next level: You prove triangles congruent and then use that knowledge to prove other things. So, proving triangles congruent can be either the final goal of a proof or a stepping stone.

TIP

Look for congruent triangles: Proving triangles congruent is critical, and thus you should always check the proof diagram and find *all* pairs of triangles that look like they have the same shape and size. If you find any, you very likely will have to prove one (or more) of the pairs of triangles congruent. And if you can see how to do that, you've probably won at least half the battle.

THEOREMS & POSTULATES

Okay. So here are the first three of five ways of proving two triangles congruent. (I cover the other two ways in the aptly titled section, "Two More Ways to Prove Triangles Congruent," later in the chapter. I don't give you all five at once because I don't want you to blow a geometry fuse from theorem overload.) You're going to use the five triangle theorems all the time.

>> **SSS (Side-Side-Side):** If the three sides of one triangle are congruent to the three sides of another triangle, then the triangles are congruent.

>> **SAS (Side-Angle-Side):** If two sides and the included angle of one triangle are congruent to two sides and the included angle of another triangle, then the triangles are congruent. (The *included angle* is the mathematician's fancy-pants way of saying "the angle between them.")

>> **ASA (Angle-Side-Angle):** If two angles and the included side of one triangle are congruent to two angles and the included side of another triangle, then the triangles are congruent.

And here's one more postulate that comes in handy when trying to prove triangles congruent (this wins first prize in the *well-duh* category).

THEOREMS & POSTULATES

Reflexive Property: Any segment or angle is congruent to itself. Amazing!

EXAMPLE

Q. Given: $\triangle ABC$ is isosceles with base $\overline{AC}$ and median $\overline{BM}$

Prove: $\triangle ABM \cong \triangle CBM$

A.

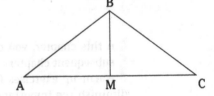

Statements	Reasons
1) $\triangle ABC$ is isosceles with base $\overline{AC}$	1) Given.
2) $\overline{AB} \cong \overline{CB}$	2) Definition of isosceles triangle.
3) $\overline{BM}$ is a median	3) Given.
4) M is the midpoint of $\overline{AC}$	4) Definition of median.
5) $\overline{AM} \cong \overline{CM}$	5) Definition of midpoint.
6) $\overline{BM} \cong \overline{BM}$	6) Reflexive Property.
7) $\triangle ABM \cong \triangle CBM$	7) SSS (2, 5, 6).

Note that after SSS in the final step, I indicate the three lines from the *statement* column where the three pairs of sides are shown to be congruent. Doing so is optional, but it's a good idea, because it can help you avoid some careless mistakes. *Remember:* The three lines you list must show three congruencies of segments or angles (three pairs of congruent segments in the current problem).

Q. Given: *B* and *C* trisect $\overline{AD}$

                 $\angle 1 \cong \angle 2$

                 $\triangle BCE$ is isosceles with base $\overline{BC}$

EXAMPLE   Prove: $\triangle ABE \cong \triangle DCE$

A.

Statements	Reasons
1) *B* and *C* trisect $\overline{AD}$	1) Given.
2) $\overline{AB} \cong \overline{DC}$	2) If a segment is trisected, then it is divided into three congruent segments (definition of trisect).
3) $\angle 1 \cong \angle 2$	3) Given.
4) $\angle ABE$ is supplementary to $\angle 1$ $\angle DCE$ is supplementary to $\angle 2$	4) If two angles form a straight angle (assumed from diagram), then they are supplementary.
5) $\angle ABE \cong \angle DCE$	5) If two angles are congruent, then their supplements are congruent.
6) $\triangle BCE$ is isosceles with base $\overline{BC}$	6) Given.
7) $\overline{BE} \cong \overline{CE}$	7) Definition of isosceles triangle.
8) $\triangle ABE \cong \triangle DCE$	8) SAS (2, 5, 7).

Q. Given: $\overline{PS}$ is an altitude of $\triangle QPT$

                 $\overline{PS}$ bisects $\angle RPT$

                 $\overline{PR}$ bisects $\angle QPS$

EXAMPLE $\overline{PT}$ bisects $\angle UPS$

     Prove: $\triangle QPS \cong \triangle UPS$

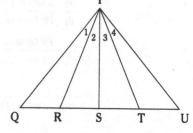

A.   **Game plan:** $\overline{PS}$ is an altitude, so it's perpendicular to the base, and the perpendicularity gives you congruent right angles *PSQ* and *PSU* — one congruence down, two to go. The two triangles share $\overline{PS}$ — two down, one to go.

So far, you have a pair of congruent angles and a pair of congruent sides, and all the rest of the givens concern the angles near *P*, so this is almost certainly an ASA problem. All that's left is to show that ∠*QPS* is congruent to ∠*UPS*. Because of the first bisection, ∠2 ≅ ∠3 (say they're each 20°). Finally, because of the other two bisections, ∠*QPS* is twice as big as ∠2 and ∠*UPS* is twice ∠3, so ∠*QPS* has to be congruent to ∠*UPS* (they'd both be 40°). Seems simple, huh?

Statements	Reasons
1) $\overline{PS}$ is an altitude of △*QPT*	1) Given.
2) $\overline{PS} \perp \overline{QT}$	2) Definition of altitude.
3) ∠*PSQ* is a right angle ∠*PSU* is a right angle	3) Definition of perpendicular.
4) ∠*PSQ* ≅ ∠*PSU*	4) All right angles are congruent.

Four steps probably seem like a lot just to arrive at these two congruent right angles, because as soon as you see an altitude of a triangle, you know you've got two congruent angles. But that's the way proofs work. You have to put down every link in the chain of logic — even incredibly obvious ones. (Like Step 2, for example: You can't just jump from Step 1 to Step 3 even though Step 3 is obvious when you know Step 1.) Every little step must be spelled out — sort of like if you had to make the logic understandable to a computer. And here's how a computer "thinks":

If Altitude then Perpendicular

If Perpendicular then Right angles

If Right angles then Congruent

In short, A → P; P → R; R → C. You need this complete chain of logic. Like it or not, those are the rules of the game. And if you're going to play the proof game, you've got to play by the rules.

5) $\overline{PS} \cong \overline{PS}$	5)	Reflexive Property. (Well, that was easy.)
6) $\overline{PS}$ bisects ∠*RPT*	6)	Given.
7) ∠*RPS* ≅ ∠*TPS*	7)	Definition of bisect.
8) $\overline{PR}$ bisects ∠*QPS* $\overline{PR}$ bisects ∠*UPS*	8)	Given.
9) ∠*QPS* ≅ ∠*UPS*	9)	If two angles are congruent (∠*RPS* and ∠*TPS*), then their like multiples are congruent (∠*QPS* is double ∠*RPS* and ∠*UPS* is double ∠*TPS*).
10) △*QPS* ≅ △*UPS*	10) ASA (4, 5, 9).

1. Given: $\overline{AE}$ is an altitude and a median

   Prove: $\triangle LEA \cong \triangle MEA$

Statements	Reasons

2. Given: $\overline{SQ}$ bisects $\overline{PT}$

   $\angle 1 \cong \angle 2$

   Prove: $\triangle PQR \cong \triangle TSR$

Statements	Reasons

3 Given: $\triangle TAG$ is isosceles with base $\overline{TG}$

$\overline{TH} \cong \overline{GN}$

Prove: $\triangle TAN \cong \triangle GAH$

Hint: If this proof has you flummoxed or flabbergasted, try adding just Statements 5 and 6 from the solution.

Statements	Reasons

4 Given: $\angle AMN$ is complementary to $\angle TAX$

$\angle ATX$ is complementary to $\angle MAN$

A is the midpoint of $\overline{TM}$

Prove: $\triangle TAX \cong \triangle MAN$

Statements	Reasons

*5 Given: $\triangle IML$ and $\triangle DRO$ are isosceles with bases $\overline{IL}$ and $\overline{DO}$

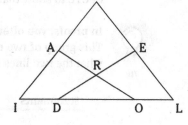

A is the midpoint of $\overline{IM}$

E is the midpoint of $\overline{LM}$

R is the midpoint of $\overline{AO}$ and $\overline{ED}$

$\overline{ID} \cong \overline{LO}$

Prove: $\triangle IAO \cong \triangle LED$

Hint: If you get stuck, copy just Statement 4 and Reason 4 from the solution and try again. If you still need a boost, you can copy Statement 8 and Reason 8 as well.

Statements	Reasons

Corresponding Parts of Congruent Triangles Are Congruent

Contrary to popular belief, CPCTC does not stand for *Cows Pull Carts To China*; it's the acronym for *Corresponding Parts of Congruent Triangles are Congruent*.

CPCTC: If two triangles are congruent, then their corresponding parts are congruent.

REMEMBER Here's how you use CPCTC. In a proof, whenever you prove two triangles congruent, you'll use CPCTC on the very next line as the justification for stating that two sides or two angles (of the two triangles) are congruent. Every triangle has six parts: three sides and three angles. You need to use three out of the six parts when you prove two triangles congruent with SSS, SAS, or ASA (see the preceding section for more on these methods of showing congruency). Therefore, there are always three other pairs of sides or angles that you haven't used yet, and you'll use CPCTC to show that one of those pairs is congruent.

TIP

In proofs, you often prove two triangles congruent and then use CPCTC on the following line. This group of two consecutive lines makes up the core or heart of many, many proofs. Here's what the two lines might look like:

Statements	Reasons
. . .	. . .
. . .	. . .
. . .	. . .
7) $\triangle ABC \cong \triangle DEF$	7) SAS.
8) $\overline{BC} \cong \overline{EF}$	8) CPCTC.
. . .	. . .
. . .	. . .
. . .	. . .

In Chapter 2, I tell you to make sure you use every given. Thinking about how to use the givens is essential at the beginning of a proof. Then I give you a tip about working backwards from the end of a proof. Both are great strategies for solving proofs. The preceding tip about using CPCTC right after showing triangles to be congruent is sort of about working at the *middle* of a proof. The key to many proofs is a pair of lines like those two lines 7 and 8. If you attack proofs like this at their beginning, middle, and end, even the longest, gnarliest proofs won't stand a chance.

For the upcoming CPCTC example proof, the diagram and the givens are identical to those from the first example in the section, "Sizing Up Three Ways to Prove Triangles Congruent." Only the *prove* statement is different.

EXAMPLE

Q.   Given: $\triangle ABC$ is isosceles with base $\overline{AC}$ and median $\overline{BM}$

Prove: $\overline{BM}$ bisects $\angle ABC$

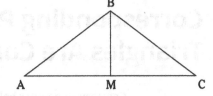

A. This proof is not particularly long or difficult, but — especially if you haven't seen the similar problem in the previous section — it may look just a bit tricky at first glance (you may wonder what the *prove* statement has to do with the givens). Look how easy the proof becomes when you work backwards.

Game plan: Start at the end. You have to prove that $\overrightarrow{BM}$ bisects $\angle ABC$. The only way to do that is to use the definition of bisect; in other words, you have to show $\angle ABM \cong \angle CBM$. Those angles are in triangles that look congruent: $\triangle ABM$ and $\triangle CBM$. Therefore, if you can show $\triangle ABM \cong \triangle CBM$, you can get $\angle ABM \cong \angle CBM$ with CPCTC. Okay — so if you can prove the triangles congruent, you're home free. At this point, you would go to the beginning of the proof and try to figure out how to show that the triangles are congruent. But here you can cut to the chase, because you already know how to get the triangles congruent from the proof in the preceding section.

Statements	Reasons
1) $\triangle ABC$ is isosceles with base $\overline{AC}$	1) Given.
2) $\overline{AB} \cong \overline{CB}$	2) Definition of isosceles triangle.
3) $\overline{BM} \cong \overline{BM}$	3) Reflexive Property.
4) $\overline{BM}$ is a median	4) Given.
5) M is the midpoint of $\overline{AC}$	5) Definition of median.
6) $\overline{AM} \cong \overline{CM}$	6) Definition of midpoint.
7) $\triangle ABM \cong \triangle CBM$	7) SSS (2, 3, 6).
8) $\angle ABM \cong \angle CBM$	8) CPCTC.
9) $\overrightarrow{BM}$ bisects $\angle ABC$	9) Definition of bisect.
	Or, if you feel like practicing your if-then logic: "If a ray divides an angle into two congruent angles (Statement 8), then it bisects the angle (Statement 9)."

6 Given: $\overline{AE}$ is an altitude and $\overline{AE}$ bisects $\angle LAM$

Prove: $\overline{AE}$ is a median

Hint: This proof is quite similar to problem 1 in the preceding section, but try to do it without looking back. If you really need something to go on, however, copy only Statement 8 and Reason 9 from the solution.

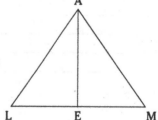

Statements	Reasons

7 Given: △OXE is isosceles with base $\overline{OE}$

$\angle XOE \cong \angle XEO$

$\angle DXE \cong \angle RXO$

Prove: $\overline{DO} \cong \overline{RE}$

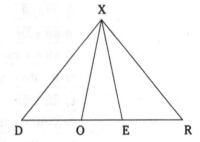

You have two different ways to do this proof. *Hint:* The shorter proof goes one step beyond CPCTC; the longer proof uses CPCTC as the final reason. (I realize that sounds like I mixed up "shorter" and "longer," but you can see in a minute that it's correct.)

Statements	Reasons

8 Given: $\overline{ZA} \cong \overline{XI}$

$\angle MZA \cong \angle FXI$

$\angle FAZ \cong \angle MIX$

Prove: $\overline{MI} \cong \overline{FA}$

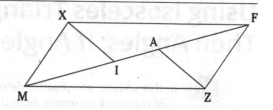

Hint: If this proof freaks you out, fill in Statements 5 and 7 and Reason 5 from the solution.

Statements	Reasons

*9 Given: $\triangle TAG$ is isosceles with base $\overline{TG}$

$\overline{TH} \cong \overline{GN}$

Prove: $\triangle THX \cong \triangle GNX$

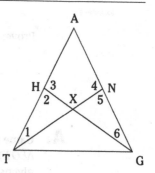

Does this *thang* ring a bell? This proof is the same as problem 3 except that it goes further. But don't look back unless you need a hint. You could also copy Statements 6 and 7 from the solution page with their Reasons; and if that's not enough, copy Statement 10 and its Reason as well.

Statements	Reasons

Using Isosceles Triangle Rules: If Sides, Then Angles; If Angles, Then Sides

THEOREMS & POSTULATES

In this section, you practice doing problems involving two of the most important and often-used theorems for proofs. Both theorems are about isosceles triangles. But these theorems are really just a single idea that works in both directions.

>> **If sides, then angles:** If two sides of a triangle are congruent, then the angles opposite those sides are congruent.

>> **If angles, then sides:** If two angles of a triangle are congruent, then the sides opposite those angles are congruent.

TIP

Look for isosceles triangles. These two angle-side theorems come up all the time in proofs. So, when you begin a proof, look at the diagram and identify all triangles that look isosceles. Make a mental note that you may have to use one or the other of the theorems for one or more of the isosceles triangles. Because recognizing isosceles triangles is often a cinch — and because it's so easy to use the theorems — be glad when you get these "gimmes" in a proof. On the other hand, if you fail to notice that the theorems should be used, the proof may become impossible. Forewarned is forearmed.

EXAMPLE

Q. Given: $\overline{OE}$ and $\overline{OT}$ trisect $\overline{NW}$

           $\overline{ON} \cong \overline{OW}$

   Prove: $\triangle ONE \cong \triangle OWT$

A.   **Game plan:** First, you look at the diagram and see two isosceles triangles ($\triangle NOW$ and $\triangle EOT$), so you're rarin' to use one of the angle-side theorems. Sure enough, one of the givens is $\overline{ON} \cong \overline{OW}$, so that gives you $\angle N \cong \angle W$. The trisection gives you $\overline{NE} \cong \overline{WT}$, and, voilà, you have SAS.

Statements	Reasons
1) $\overline{ON} \cong \overline{OW}$	1) Given.
2) $\angle N \cong \angle W$	2) If sides, then angles.
3) $\overline{OE}$ and $\overline{OT}$ trisect $\overline{NW}$	3) Given.
4) $\overline{NE} \cong \overline{WT}$	4) Definition of segment trisection.
5) $\triangle ONE \cong \triangle OWT$	5) SAS (1, 2, 4).

Q. Given: $\angle A \cong \angle E$

$\overline{GM}$ is a median

EXAMPLE

Prove: $\triangle MAG \cong \triangle MEG$

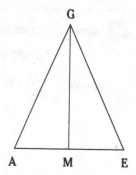

A.

Statements	Reasons
1) $\angle A \cong \angle E$	1) Given.
2) $\overline{AG} \cong \overline{EG}$	2) If angles, then sides. (***Repeat tip:*** Do not fail to spot this!)
3) $\overline{GM}$ is a median	3) Given.
4) M is the midpoint of $\overline{AE}$	4) Definition of median.
5) $\overline{AM} \cong \overline{EM}$	5) Definition of midpoint.
6) $\triangle MAG \cong \triangle MEG$	6) SAS (2, 1, 5).
	(Note that if you add one more step for $\overline{GM}$ reflexive, you can finish with SSS instead of SAS. This six-step solution is a fairly unusual proof where you have side-by-side triangles like this but don't use the Reflexive Property.)

10 Given: $\angle 1 \cong \angle 2$

$\angle 3 \cong \angle 4$

Prove: $\triangle RAY \cong \triangle BAN$

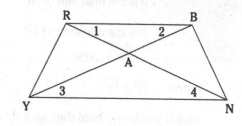

Statements	Reasons

11 Given: $\triangle QRS$ is isosceles with base $\overline{QS}$

$\angle QPT \cong \angle STP$

Prove: $\overline{PQ} \cong \overline{TS}$

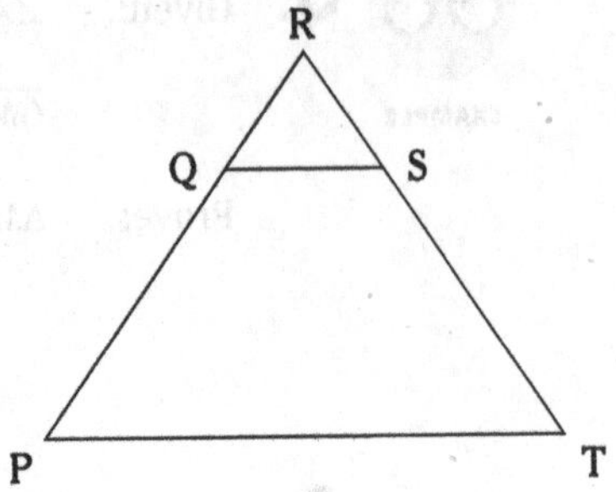

Statements	Reasons

12 Given: $\overline{AB} \cong \overline{ED}$

B is the midpoint of $\overline{AC}$

D is the midpoint of $\overline{EC}$

$\angle ABF \cong \angle EDF$

Prove: $\overline{BF} \cong \overline{DF}$

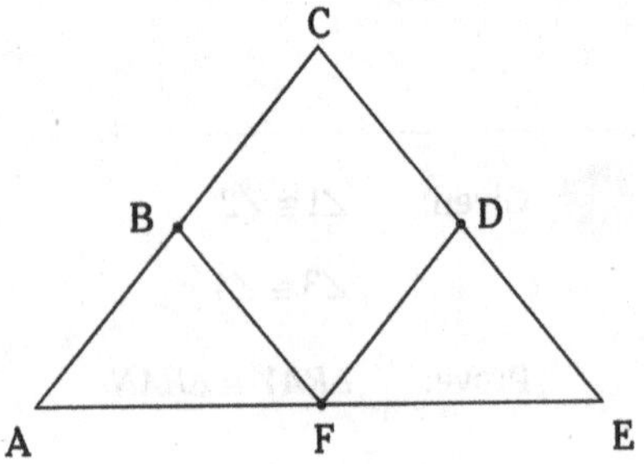

Hint: If you have a hard time with this one, copy Statement 3 and Reason 3 from the solution. If you're still stuck, copy Statement 7 and Reason 7 as well.

Statements	Reasons

Exploring Two More Ways to Prove Triangles Congruent

THEOREMS & POSTULATES

You have five ways of showing triangles congruent. The previous sections in this chapter let you practice problems with SSS, SAS, and ASA. Now you get the final two methods, AAS and HL.

>> **AAS (Angle-Angle-Side):** If two angles and a nonincluded side of one triangle are congruent to the corresponding parts of another triangle, then the triangles are congruent.

>> **HL (Hypotenuse-Leg):** If the hypotenuse and a leg of one right triangle are congruent to the hypotenuse and a leg of another right triangle, then the triangles are congruent.

AAS works in nearly the same way as SSS, SAS, and ASA. If two triangles have congruent angles, then congruent angles, then congruent sides (in that order, going around the triangles clockwise or counterclockwise), then the triangles are congruent. Like SSS, SAS, and ASA, AAS works with any type of triangle.

WARNING

ASS or ASS-backwards is no good. You can prove triangles congruent with SSS, SAS, ASA, and AAS, but *not* with ASS or SSA. SAA would work (but you just call it AAS). In short, every three-letter combination of *A*s and *S*s works unless it spells *ass* or is *ass*-backwards (SSA). (AAA — which you get to in Chapter 7 — also "works," but not to show that triangles are congruent. You use it to show that triangles are similar.)

HL is a bit different from the other four theorems because it works only with right triangles. For this reason, if I were writing my own book, I'd add the letter *R* and call it HLR (for Hypotenuse, Leg, Right angle). Wait a minute — I *am* writing my own book! Okay, so contrary to other books, I will call it HLR. (Please go to the section intro and the theorem icon, scratch out HL, and replace it with HLR.) HLR is a better name because its three letters make you focus on the fact that when you use HLR — just like with SSS, SAS, ASA, and AAS — you need *three* things to prove two triangles congruent.

Note that, in terms of *A*s and *S*s, HLR is *ass*-backwards (SSA), because going around the triangle you use a *Side* (the Hypotenuse), a *Side* (the Leg), and an *Angle* (the Right angle) in that order. Thus, HLR is a valid, special case of SSA and an exception to the general invalidity of SSA.

Before going on to the problems, I have one more theorem for you. It's yet another in the *well-duh* category. I don't use it in the following example problems, but you do need it for your practice problems.

THEOREMS & POSTULATES

Congruent plus supplementary means right angles: If two angles are both congruent and supplementary, then they're right angles.

Q. Given: ∠RTP ≅ ∠RPT

∠PQR ≅ ∠TSR

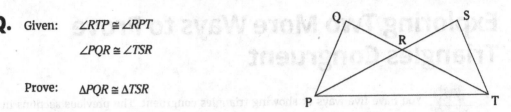

Prove: △PQR ≅ △TSR

A.

Statements	Reasons
1) ∠RTP ≅ ∠RPT	1) Given.
2) $\overline{PR}$ ≅ $\overline{TR}$	2) If angles, then sides.
3) ∠PRQ ≅ ∠TRS	3) Vertical angles are congruent.
4) ∠PQR ≅ ∠TSR	4) Given.
5) △PQR ≅ △TSR	5) AAS (4, 3, 2).

Q. Given: $\overline{RS} \perp \overline{RO}$

$\overline{AE} \perp \overline{AN}$

$\overline{OE} \cong \overline{NS}$

$\overline{RS} \cong \overline{AE}$

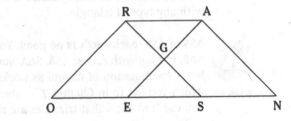

Prove: △ROS ≅ △ANE

A.

Statements	Reasons
1) $\overline{RS} \perp \overline{RO}$ $\overline{AE} \perp \overline{AN}$	1) Given.
2) ∠ORS is a right angle ∠NAE is a right angle	2) Definition of perpendicular.
3) $\overline{OE} \cong \overline{NS}$	3) Given.
4) $\overline{OS} \cong \overline{NE}$	4) Segment addition. (If a segment [$\overline{ES}$] is added to two congruent segments [$\overline{OE}$ and $\overline{NS}$], then the sums [$\overline{OS}$ and $\overline{NE}$] are congruent.)
5) $\overline{RS} \cong \overline{AE}$	5) Given.
6) △ROS ≅ △ANE	6) HLR (4, 5, 2). (Note that for HLR, you need to state only that you have two right angles, not that they are congruent.)

13 Given: $\overline{IN} \cong \overline{IO}$

   $\angle OWN \cong \angle NKO$

Prove: $\triangle WOZ \cong \triangle KNZ$

Hint: If you get stuck, copy just Statements 5 and 6 with their Reasons onto this page.

Statements	Reasons

*14 Given: $\angle TIN$ and $\angle EAR$ are right angles

   $\overline{AX}$ bisects $\overline{TI}$

   $\overline{IZ}$ bisects $\overline{EA}$

   $\overline{AX}$ and $\overline{IZ}$ trisect $\overline{TE}$

   $\overline{XI} \cong \overline{ZE}$

Prove: $\overline{IN} \cong \overline{AR}$

Statements	Reasons

*15 Given: $\overline{AB} \cong \overline{CD}$

$\angle BFA \cong \angle BFE$

$\angle DEC \cong \angle DEF$

$\overline{BC} \perp \overline{AB}$

$\overline{AD} \perp \overline{DC}$

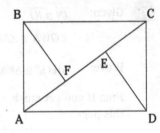

Prove: $\overline{FB} \cong \overline{ED}$

Hint: If you're stumped, write in Statements and Reasons 5 and 6 from the solution.

Statements	Reasons

Explaining the Two Equidistance Theorems

Throughout this chapter, I emphasize how important it is to pay attention to the congruent triangles in proof diagrams because the key to so many proofs is showing the triangles congruent and then using CPCTC. Now I muddy the waters a bit by giving you two theorems that you can often use *instead* of proving triangles congruent. You may get proofs in which you see congruent triangles, so it looks like you should try to show that the triangles are congruent, but you don't have to — one of the *equidistance* theorems can give you a shortcut to the final conclusion.

Now you have to be doubly on your toes: looking for congruent triangles and thinking about ways to prove them congruent and, at the same time, being ready to avoid the congruent triangle issue with the equidistance shortcut.

THEOREMS &
POSTULATES

The equidistance theorems:

>> If two points are each (one at a time) equidistant from the endpoints of a segment, then those points determine the perpendicular bisector of the segment. (Loose, short form: If *two* pairs of congruent segments, then perpendicular bisector.)

>> If a point is on the perpendicular bisector of a segment, then it's equidistant from the endpoints of the segment. (Loose, short form: If perpendicular bisector, then *one* pair of congruent segments.)

These theorems are a royal mouthful. The best way to understand them is visually. For the first theorem, consider Figure 5-1.

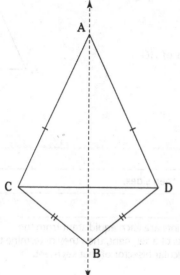

FIGURE 5-1:
The first
equidistance
theorem
gives you the
perpendicular
bisector of a
segment.

Here's how the theorem works. If you have one point (like *A*) that's equally distant from the endpoints of a segment ($\overline{CD}$) and another point (like *B*) that's also equally distant from those endpoints, then the two points (*A* and *B*) determine (show you where to draw) the perpendicular bisector of that segment. The dashed line in the figure, $\overleftrightarrow{AB}$, is the perpendicular bisector of $\overline{CD}$, which means — as I'm sure you know or can figure out — that it's perpendicular to $\overline{CD}$ and cuts $\overline{CD}$ in half. I didn't mark the perpendicularity or the bisection in the figure because I wanted to mark only the *if* part of the theorem. The figure should also make clear the meaning of the loose, short form of the theorem: If *two* pairs of congruent segments ($\overline{AC} \cong \overline{AD}$ and $\overline{BC} \cong \overline{BD}$), then perpendicular bisector ($\overleftrightarrow{AB}$ is the perpendicular bisector of $\overline{CD}$).

For the second theorem, consider Figure 5-2.

The second theorem tells you that if you start with a segment (like $\overline{PQ}$) and its perpendicular bisector (like line *l*), and a point is on the perpendicular bisector (like *R*), then *R* is equally distant from the endpoints of the segment. The figure also illustrates the loose, short form of the theorem: If perpendicular bisector (line *l* is the perpendicular bisector of $\overline{PQ}$), then *one* pair of congruent segments ($\overline{RP} \cong \overline{RQ}$).

FIGURE 5-2:
The second equidistance theorem lets you know that you have congruent segments.

Q. Given: $\overline{TA} \cong \overline{TO}$

$\angle PAD \cong \angle POD$

Prove: $\overleftrightarrow{PT}$ is the perpendicular bisector of $\overline{AO}$

EXAMPLE

A.

Statements	Reasons
1) $\angle PAD \cong \angle POD$	1) Given.
2) $\overline{PA} \cong \overline{PO}$	2) If angles, then sides.
3) $\overline{TA} \cong \overline{TO}$	3) Given.
4) $\overleftrightarrow{PT}$ is the perpendicular bisector of $\overline{AO}$	4) If two points are each equidistant from the endpoints of a segment, then they determine the perpendicular bisector of that segment.

Q. Given: $\triangle TIC$ and $\triangle TOC$ are isosceles triangles with base $\overline{TC}$

EXAMPLE Prove: $\triangle TAC$ is isosceles

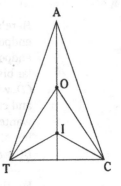

A.

Statements	Reasons
1) △*TIC* and △*TOC* are isosceles	1) Given.
2) $\overline{TI} \cong \overline{CI}$ $\overline{TO} \cong \overline{CO}$	2) Definition of isosceles triangle.
3) $\overleftrightarrow{OI}$ is the perpendicular bisector of $\overline{TC}$	3) If two points are each equidistant from the endpoints of a segment, then they determine the perpendicular bisector of that segment.
4) $\overline{TA} \cong \overline{CA}$	4) If a point is on the perpendicular bisector of a segment, then it is equidistant from the endpoints of that segment.
5) △*TAC* is isosceles	5) Definition of isosceles triangle.

Tip: When you see an unlabeled point in a problem, you don't need to use that point in your proof. Note that in the preceding figure, no letter labels the point where the perpendicular bisector $\overleftrightarrow{OI}$ intersects $\overline{TC}$. This tip doesn't help so much in this particular problem, but for some proofs, this built-in hint can work wonders.

---

16 Given: $\angle TIP \cong \angle TOP$

$\angle HIP \cong \angle HOP$

Prove: $\overline{IP} \cong \overline{OP}$

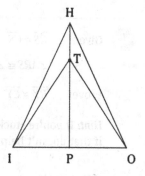

Statements	Reasons

17 Given: ∠SAT and ∠SET are right angles

SA ≅ SE

Prove: ∠ANT ≅ ∠ENT

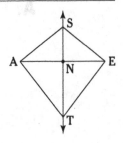

Statements	Reasons

18 Given: RS ≅ CS

∠ARS ≅ ∠ACS

Prove: RY ≅ CY

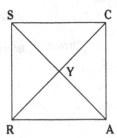

Hint: If you're stuck, copy only Statement 2 and Reason 2 from the solution. And if that doesn't help, copy Statement 5 and Reason 5 as well.

Statements	Reasons

Solutions

(1)

Statements	Reasons
1) $\overline{AE} \cong \overline{AE}$	1) Reflexive.
2) $\overline{AE}$ is an altitude	2) Given.
3) $\overline{AE} \perp \overline{LM}$	3) Definition of altitude. (Why would they tell you about an altitude? Because it's perpendicular.)
4) $\angle LEA$ is a right angle $\angle MEA$ is a right angle	4) Definition of perpendicular. (What does perpendicularity tell you? Right angles, of course.)
5) $\angle LEA \cong \angle MEA$	5) All right angles are congruent.
6) $\overline{AE}$ is a median	6) Given.
7) E is the midpoint of $\overline{LM}$	7) Definition of median. (Why would they tell you about a median? Because it goes to a midpoint.)
8) $\overline{LE} \cong \overline{ME}$	8) Definition of midpoint. (What do you know about a midpoint? Congruent segments, of course.)
9) $\triangle LEA \cong \triangle MEA$	9) SAS (1, 5, 8).

(2)

Statements	Reasons
1) $\angle PRQ \cong \angle TRS$	1) Vertical angles are congruent. (This step should be a no-brainer. Always check for vertical angles.)
2) $\overline{SQ}$ bisects $\overline{PT}$	2) Given. (Now, why would they tell you this? Only one possible reason . . .)
3) $\overline{PR} \cong \overline{TR}$	3) Definition of bisect. (So far, you have one pair of congruent angles and one pair of congruent sides, so this problem has to end with either ASA or SAS, right? If it's ASA, you need to show $\angle QPR \cong \angle STR$. If it's SAS, you need to show $\overline{SR} \cong \overline{QR}$. Which seems more promising? It's ASA, because the remaining given concerns $\angle 1$ and $\angle 2$, which are right next to $\angle QPR$ and $\angle STR$. Don't forget: Every given is a built-in hint.)
4) $\angle 1 \cong \angle 2$	4) Given.
5) $\angle QPR \cong \angle STR$	5) Supplements of congruent angles are congruent.
6) $\triangle PQR \cong \triangle TSR$	6) ASA (1, 3, 5).

(3)

Statements	Reasons
1) $\triangle TAG$ is isosceles with base $\overline{TG}$	1) Given.
2) $\overline{TA} \cong \overline{GA}$	2) Definition of isosceles triangle.
3) $\angle A \cong \angle A$	3) Reflexive Property. (Did you notice that $\angle A$ is in both of the *prove* triangles? If not, open your eyes!)
4) $\overline{TH} \cong \overline{GN}$	4) Given.

If you're not sure where to go from here, try this technique. You know △*TAG* is isosceles, so make up a length for sides $\overline{TA}$ and $\overline{GA}$, say, 10. Then make up a length for congruent segments $\overline{TH}$ and $\overline{GN}$, say, 7. What follows? That *HA* and *NA* are both 3, of course, and therefore, that they're congruent. And that's all you need for SAS.

5) $\overline{HA} \cong \overline{NA}$	5) If congruent segments are subtracted from congruent segments, then the differences are congruent.
6) △*TAN* ≅ △*GAH*	6) SAS (2, 3, 5).

(4)

Statements	Reasons
1) ∠*TAX* ≅ ∠*MAN*	1) Vertical angles are congruent. (Always look for vertical angles!)
2) ∠*AMN* is complementary to ∠*TAX* ∠*ATX* is complementary to ∠*MAN*	2) Given.
3) ∠*AMN* ≅ ∠*ATX*	3) Complements of congruent angles are congruent. Or, if you prefer the long way, "If two angles are congruent (Statement 1), then their complements are congruent (Statement 3)."
4) *A* is the midpoint of $\overline{TM}$	4) Given.
5) $\overline{TA} \cong \overline{MA}$	5) Definition of midpoint.
6) △*TAX* ≅ △*MAN*	6) ASA (1, 5, 3).

(*5) **Game plan:** The big triangle is isosceles, so you have two equal sides. The midpoints cut those sides in half, so all four halves are equal and *IA* equals *LE*. The little triangle is also isosceles, with equal sides $\overline{DR}$ and $\overline{OR}$. Then, because *R* is a midpoint (for two segments), *DE* is twice *DR* and *OA* is twice *OR*; thus, *DE* equals *OA*. The last given is $\overline{ID} \cong \overline{LO}$. Say the segments are each 4 units long; and say *DO* is 8. That makes *IO* and *LD* both 12. That's a good bingo — SSS.

Statements	Reasons
1) △*IML* is isosceles with base $\overline{IL}$	1) Given.
2) $\overline{IM} \cong \overline{LM}$	2) Definition of isosceles triangle.
3) *A* is the midpoint of $\overline{IM}$ *E* is the midpoint of $\overline{LM}$	3) Given.
4) $\overline{IA} \cong \overline{LE}$	4) Like Divisions. Or, to make sure you're following proper if-then logic, "If two segments are congruent (Statement 2), then their like divisions are congruent (Statement 4)."
5) △*DRO* is isosceles with base $\overline{DO}$	5) Given.
6) $\overline{OR} \cong \overline{DR}$	6) Definition of isosceles triangle.
7) *R* is the midpoint of $\overline{AO}$ *R* is the midpoint of $\overline{ED}$	7) Given.

8) $\overline{OA} \cong \overline{DE}$	8) Like Multiples.
	Basically, if two segments are congruent ($\overline{OR}$ and $\overline{DR}$), then twice one ($\overline{OA}$) equals twice the other ($\overline{DE}$).
9) $\overline{ID} \cong \overline{LO}$	9) Given.
10) $\overline{IO} \cong \overline{LD}$	10) If a segment is added to two congruent segment, then the sums are congruent ($\overline{DO}$ is added to $\overline{ID}$ and $\overline{LO}$).
11) $\triangle IAO \cong \triangle LED$	11) SSS (4, 8, 10).

(6) **Game plan:** Work backwards. To prove $\overline{AE}$ is a median, you need to show $\overline{LE} \cong \overline{ME}$. And you can probably get that with CPCTC after showing that the triangles are congruent. 'Nuff said.

Statements	Reasons
1) $\overline{AE}$ is an altitude	1) Given.
2) $\overline{AE} \perp \overline{LM}$	2) Definition of altitude.
3) $\angle LEA$ is a right angle	3) Definition of perpendicular.
$\angle MEA$ is a right angle	
4) $\angle LEA \cong \angle MEA$	4) Right angles are congruent.
5) $\overline{AE} \cong \overline{AE}$	5) Reflexive.
6) $\overline{AE}$ bisects $\angle LAM$	6) Given.
7) $\angle LAE \cong \angle MAE$	7) Definition of bisect.
8) $\triangle LEA \cong \triangle MEA$	8) ASA (4, 5, 7).
9) $\overline{LE} \cong \overline{ME}$	9) CPCTC.
10) E is the midpoint of $\overline{LM}$	10) Definition of midpoint.
11) $\overline{AE}$ is a median	11) Definition of median.

(7) **Game plan (shorter version):** Always take a quick glance at proof diagrams and look for triangles that look congruent. In this diagram, you should see two such pairs: $\triangle DXO$ and $\triangle RXE$ and $\triangle DXE$ and $\triangle RXO$. The three givens basically hand you that second pair of triangles on a silver platter (with ASA). You get $\overline{DE} \cong \overline{RO}$ with CPCTC; then subtract $\overline{OE}$ and you're done.

Statements	Reasons
1) $\triangle OXE$ is isosceles with base $\overline{OE}$	1) Given.
2) $\overline{OX} \cong \overline{EX}$	2) Definition of isosceles triangle.
3) $\angle XOE \cong \angle XEO$	3) Given.
4) $\angle DXE \cong \angle RXO$	4) Given.
5) $\triangle DXE \cong \triangle RXO$	5) ASA (3, 2, 4).
6) $\overline{DE} \cong \overline{RO}$	6) CPCTC.
7) $\overline{DO} \cong \overline{RE}$	7) Segment subtraction.
	(If a segment is subtracted from two congruent segments, then the differences are congruent. Say DE and RO are both 8 and OE is 3. Then DO and RE would both be 5.)

TIP

Stay flexible. The fact that you can do this proof (and many others) in more than one way shows that geometry proofs aren't quite as cut and dried as some other types of math problems. And because of that, you should adopt a flexible approach when doing proofs. Don't assume that there's just one precise way of doing a proof and that you're sunk if you can't find it. Be flexible, use some trial and error, use your imagination. Try anything — don't worry about whether it's "right." Be patient with yourself, and don't expect to always solve a proof on your first try. You have to be willing to try something, find yourself at a dead end, and then go back to the drawing board to try something else. And although it's true that, as a general rule, the fewer steps in your proof, the better, you shouldn't worry too much about that. Most teachers don't mind if your method is a little longer than the shortest possible proof. They may, however, take off a few points if your proof is *way* longer than it has to be.

Game plan (longer version): Say you notice the other pair of congruent triangles, $\triangle DXO$ and $\triangle RXE$, and realize that the proof can, thus, end with CPCTC. You then have to find a way to show $\triangle DXO \cong \triangle RXE$. You already have $\overline{OX} \cong \overline{EX}$. Then you can say $\angle DOX \cong \angle REX$, because their supplements are congruent. Finally, you can get $\angle DXO \cong \angle RXE$ by subtracting the middle angle, $\angle OXE$, from the overlapping congruent angles, $\angle DXE$ and $\angle RXO$. Not bad, right?

Statements	Reasons
1) $\triangle OXE$ is isosceles with base $\overline{OE}$	1) Given.
2) $\overline{OX} \cong \overline{EX}$	2) Definition of isosceles triangle.
3) $\angle XOE \cong \angle XEO$	3) Given.
4) $\angle DOX \cong \angle REX$	4) Supplements of congruent angles are congruent.
5) $\angle DXE \cong \angle RXO$	5) Given.
6) $\angle DXO \cong \angle RXE$	6) Angle subtraction.
7) $\triangle DXO \cong \triangle RXE$	7) ASA (4, 2, 6).
8) $\overline{DO} \cong \overline{RE}$	8) CPCTC.

⑧ **Game plan:** You have two pairs of congruent triangles that include $\overline{ZA}$ and $\overline{XI}$. Which pair of triangles should you shoot for? Stay flexible. Look at the pair including $\angle MXI$ and $\angle FZA$. (If you can prove that pair congruent, you can finish with CPCTC.) You have $\overline{XI}$ and $\angle MIX$. Can you get the third element you need for SAS or ASA? To use SAS, you'd need to know that $\overline{MI} \cong \overline{FA}$, but that's what you're trying to prove, so that won't work. For ASA, you'd need to work your way to $\angle MXI \cong \angle FZA$, but there doesn't seem to be a way to get that. You appear to be at a dead end, so it's time to go back to the drawing board. (By the way, winding up in a dead end like this is par for the course with geometry proofs. Trying to show $\triangle MXI \cong \triangle FZA$ is a perfectly good idea. You have two out of three triangle parts that you need for something like ASA, and the triangles are a natural choice because of the possibility of finishing the proof with CPCTC. Don't let dead ends like this frustrate you.)

Time to try the second pair of triangles. Look at $\triangle MAZ$. You have $\triangle MZA$ and side $\overline{ZA}$. Can you get $\angle MAZ$ and finish with ASA? Yes. That's it. You have $\angle FAZ \cong \angle MIX$ (say, for instance, they're both 50°), so $\angle MAZ \cong \angle FIX$ (they'd both be 130°). (It's not a bad idea to sometimes actually make up an angle measure like 50° and then write it on the diagram to help you see that the two 130° angles would be congruent.) You now finish with CPCTC and segment subtraction.

Statements	Reasons
1) $\overline{ZA} \cong \overline{XI}$	1) Given.
2) $\angle MZA \cong \angle FXI$	2) Given.
3) $\angle FAZ \cong \angle MIX$	3) Given.
4) $\angle MAZ \cong \angle FIX$	4) Supplements of congruent angles are congruent.
5) $\triangle MAZ \cong \triangle FIX$	5) ASA (2, 1, 4).
6) $\overline{MA} \cong \overline{FI}$	6) CPCTC.
7) $\overline{MI} \cong \overline{FA}$	7) Subtraction.

*9 **Game plan (quickie version):** You get $\triangle TAN \cong \triangle GAH$ like you do with problem 3 in the first section. Then you get $\angle 3 \cong \angle 4$ (CPCTC), their supplements, $\angle 2 \cong \angle 5$, and then $\angle 1 \cong \angle 6$ (CPCTC). Bingo! You have ASA.

Statements	Reasons
1) $\triangle TAG$ is isosceles with base $\overline{TG}$	1) Given.
2) $\overline{TA} \cong \overline{GA}$	2) Definition of an isosceles triangle.
3) $\overline{TH} \cong \overline{GN}$	3) Given.
4) $\overline{HA} \cong \overline{NA}$	4) Subtraction.
5) $\angle A \cong \angle A$	5) Reflexive.
6) $\triangle TAN \cong \triangle GAH$	6) SAS (2, 5, 4).
7) $\angle 3 \cong \angle 4$	7) CPCTC.
8) $\angle 2 \cong \angle 5$	8) Supplements of congruent angles are congruent.
9) $\angle 1 \cong \angle 6$	9) CPCTC.
10) $\triangle THX \cong \triangle GNX$	10) ASA (8, 3, 9).

10

Statements	Reasons
1) $\angle 1 \cong \angle 2$	1) Given.
2) $\overline{RA} \cong \overline{BA}$	2) If angles, then sides.
3) $\angle 3 \cong \angle 4$	3) Given.
4) $\overline{YA} \cong \overline{NA}$	4) If angles, then sides.
5) $\angle RAY \cong \angle BAN$	5) Vertical angles are congruent.
6) $\triangle RAY \cong \triangle BAN$	6) SAS (2, 5, 4).

11 **Game plan:** As soon as you see two triangles in the diagram that look like they're isosceles and then that $\angle QPT \cong \angle STP$ is given, you should immediately realize that you have $\overline{PR} \cong \overline{TR}$. The rest is in the bag.

Statements	Reasons
1) $\angle QPT \cong \angle STP$	1) Given.
2) $\overline{PR} \cong \overline{TR}$	2) If angles, then sides.
3) $\triangle QRS$ is isosceles with base $\overline{QS}$	3) Given.
4) $\overline{QR} \cong \overline{SR}$	4) Definition of isosceles triangle.
5) $\overline{PQ} \cong \overline{TS}$	5) If two congruent segments ($\overline{QR}$ and $\overline{SR}$) are subtracted from two other congruent segments ($\overline{PR}$ and $\overline{TR}$), then the differences ($\overline{PQ}$ and $\overline{TS}$) are congruent. (If PR and TR were both 10, and QR and SR were both 3, then PQ and TS would both be 7, right?)

12 **Game plan:** Maybe I sound like a broken record, but make sure you notice the three triangles that look isosceles and that, therefore, you likely have to use one of the angle-side theorems. You should also, of course, notice that the two small triangles look congruent and that the proof therefore probably ends with CPCTC. So, your goal is to show that the two triangles are congruent. You could work backwards — you already have a pair of congruent sides and a pair of congruent angles in those triangles. To finish with SAS, you'd need to use $\overline{BF} \cong \overline{DF}$, but that's what you're trying to prove. Your other option is ASA, and for that to work, you'd need $\angle A \cong \angle E$. Can you get that? This should be the light-bulb-going-on moment of the proof. You should be thinking, "I could get angle A congruent to angle E by *if sides, then angles* if I knew that $\overline{AC}$ was congruent to $\overline{EC}$." Then you can go back to the givens and see that you can, in fact, show that congruency with the Like Multiples Theorem (see Chapter 3).

Another, equally good approach to the proof is to begin by looking at the givens. You see that $\overline{AB}$ and $\overline{ED}$ are congruent and that B and D are midpoints. Now, why would they tell you about those midpoints? Midpoints cut segments in half, of course. So, if $\overline{AB} \cong \overline{ED}$, then twice $\overline{AB}$ (that's $\overline{AC}$) equals twice $\overline{ED}$ (that's $\overline{EC}$). Then, naturally, you go from $\overline{AC} \cong \overline{EC}$ to $\angle A \cong \angle E$, and the rest is a cake walk.

Statements	Reasons
1) $\overline{AB} \cong \overline{ED}$	1) Given.
2) B is the midpoint of $\overline{AC}$ D is the midpoint of $\overline{EC}$	2) Given.
3) $\overline{AC} \cong \overline{EC}$	3) Like Multiples. (If segments are congruent [$\overline{AB}$ and $\overline{ED}$], then twice one [$\overline{AC}$] is congruent to twice the other [$\overline{EC}$].)
4) $\angle A \cong \angle E$	4) If sides, then angles.
5) $\angle ABF \cong \angle EDF$	5) Given.
6) $\triangle ABF \cong \triangle EDF$	6) ASA (4, 1, 5).
7) $\overline{BF} \cong \overline{DF}$	7) CPCTC.

(13)

Statements	Reasons
1) $\overline{IN} \cong \overline{IO}$	1) Given.
2) $\angle ION \cong \angle INO$	2) If sides, then angles.
	(If you missed this, hand over your protractor! This proof is not totally easy, but this step should be.)
3) $\overline{ON} \cong \overline{NO}$	3) Reflexive.
	(Do you see why I reversed the letters?)
4) $\angle OWN \cong \angle NKO$	4) Given.
5) $\triangle WON \cong \triangle KNO$	5) AAS (4, 2, 3).
6) $\overline{OW} \cong \overline{NK}$	6) CPCTC.
7) $\angle WZO \cong \angle KZN$	7) Vertical angles are congruent.
8) $\triangle WOZ \cong \triangle KNZ$	8) AAS (7, 4, 6).

(*14)

Statements	Reasons
1) $\angle TIN$ and $\angle EAR$ are right angles	1) Given.
2) $\overline{XI} \cong \overline{ZE}$	2) Given.
3) $\overline{AX}$ bisects $\overline{TI}$	3) Given.
$\overline{IZ}$ bisects $\overline{EA}$	
4) $\overline{TI} \cong \overline{EA}$	4) Like Multiples.
	(If segments are congruent [$\overline{XI}$ and $\overline{ZE}$], then twice one [$\overline{TI}$] is congruent to twice the other [$\overline{EA}$].)
5) $\overline{AX}$ and $\overline{IZ}$ trisect $\overline{TE}$	5) Given.
6) $\overline{TR} \cong \overline{RN} \cong \overline{NE}$	6) Definition of trisect.
7) $\overline{TN} \cong \overline{ER}$	7) Segment addition.
8) $\triangle TIN \cong \triangle EAR$	8) HLR (7, 4, 1).
9) $\overline{IN} \cong \overline{AR}$	9) CPCTC.

(*15)

Statements	Reasons
1) $\overline{BC} \perp \overline{AB}$	1) Given.
$\overline{AD} \perp \overline{DC}$	
2) $\angle ABC$ is a right angle	2) Definition of perpendicular.
$\angle CDA$ is a right angle	
3) $\overline{AB} \cong \overline{CD}$	3) Given.
4) $\overline{AC} \cong \overline{CA}$	4) Reflexive.
5) $\triangle ABC \cong \triangle CDA$	5) HLR (4, 3, 2).
6) $\angle BAC \cong \angle DCA$	6) CPCTC.
7) $\angle BFA \cong \angle BFE$	7) Given.
8) $\angle BFA$ and $\angle BFE$ are right angles	8) If two angles are both congruent and supplementary, then they are right angles.

9) ∠DEC ≅ ∠DEF	9) Given.
10) ∠DEC and ∠DEF are right angles	10) If two angles are both congruent and supplementary, then they are right angles.
11) ∠BFA ≅ ∠DEC	11) All right angles are congruent.
12) △BFA ≅ △DEC	12) AAS (11, 6, 3).
13) $\overline{FB}$ ≅ $\overline{ED}$	13) CPCTC.

16

Statements	Reasons
1) ∠TIP ≅ ∠TOP	1) Given.
2) $\overline{TI}$ ≅ $\overline{TO}$	2) If angles, then sides.
3) ∠HIP ≅ ∠HOP	3) Given.
4) $\overline{HI}$ ≅ $\overline{HO}$	4) If angles, then sides.
5) $\overleftrightarrow{HT}$ is the perpendicular bisector of $\overline{IO}$	5) If two points are each equidistant from the endpoints of a segment, then they determine the perpendicular bisector of that segment.
6) $\overline{IP}$ ≅ $\overline{OP}$	6) Definition of bisect.

17

Statements	Reasons
1) ∠SAT and ∠SET are right angles	1) Given.
2) $\overline{SA}$ ≅ $\overline{SE}$	2) Given.
3) $\overline{ST}$ ≅ $\overline{ST}$	3) Reflexive.
4) △SAT ≅ △SET	4) HLR (3, 2, 1).
5) $\overline{AT}$ ≅ $\overline{ET}$	5) CPCTC.
6) $\overleftrightarrow{ST}$ is the perpendicular bisector of $\overline{AE}$	6) If two points are each equidistant from the endpoints of a segment, then they determine the perpendicular bisector of that segment.
7) ∠ANT and ∠ENT are right angles	7) Definition of perpendicular.
8) ∠ANT ≅ ∠ENT	8) All right angles are congruent.

18

Statements	Reasons
1) $\overline{RS}$ ≅ $\overline{CS}$	1) Given.
2) ∠SRY ≅ ∠SCY	2) If sides, then angles.
3) ∠ARS ≅ ∠ACS	3) Given.
4) ∠ARY ≅ ∠ACY	4) Angle subtraction.
5) $\overline{RA}$ ≅ $\overline{CA}$	5) If angles, then sides.
6) $\overleftrightarrow{SA}$ is the perpendicular bisector of $\overline{RC}$	6) If two points are each equidistant from the endpoints of a segment, then they determine the perpendicular bisector of that segment.
7) $\overline{RY}$ ≅ $\overline{CY}$	7) Definition of bisect.

3

Polygons, Proof and Non-Proof Problems

Practice problems involving quadrilaterals, pentagons, hexagons, and more.

Study the properties of the different quadrilaterals, and you find out how to prove that a four-sided figure qualifies as a particular type of quadrilateral.

Discover how to do cool things like compute the area of a polygon, the number of its diagonals, and the sum of its interior angles.

Solve problems involving *similar* polygons — that is, polygons of the exact same shape but of different sizes.

Chapter **6**

Quadrilaterals: Your Fine, Four-Sided Friends (Including Proofs)

I f you've mastered three-sided figures, you're ready to move up to four-sided figures — *quadrilaterals*. In this chapter, I tell you the defining characteristics of squares, rectangles, and kites, and I give you some pretty tidy definitions of parallelograms, rhombuses, and trapezoids as well. I also explain the properties of these different figures. Finally, I show you how to use the properties to prove that a figure is a certain type of quadrilateral — sort of like "if it walks like a duck and it quacks like a duck. . . ."

Before moving on to quadrilaterals, take a look at some important parallel line concepts that come in handy for parallelogram problems among other things.

Double-Crossers: Transversals and Their Parallel Lines

Take a look at Figure 6-1, which contains two parallel lines, the line that crosses over them (called a *transversal*), and the eight angles. Whenever you have such a situation, the following terminology applies.

REMEMBER

Angles formed by parallel lines and a transversal:

>> The pair of angles 1 and 5 (also 2 and 6, 3 and 7, and 4 and 8) are called *corresponding angles.*

>> The pair of angles 3 and 6 (as well as 4 and 5) are *alternate interior angles.*

>> Angles 1 and 8 (and angles 2 and 7) are called *alternate exterior angles.*

>> Angles 3 and 5 (and 4 and 6) are *same-side interior angles.*

>> Angles 1 and 7 (and 2 and 8) are *same-side exterior angles.*

Now, although knowing all this fancy terminology is nice, and although you need it for the following theorems (not to mention that little matter of your teacher's testing you on these terms), there's a simpler way to summarize everything you need to know about Figure 6-1.

FIGURE 6-1:
Parallel lines
and a
transversal —
angles, angles
everywhere
with lots of
them to link.

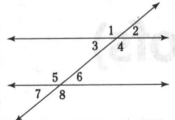

REMEMBER

Four small angles and four big angles. When you have two parallel lines cut by a transversal, you get four acute angles and four obtuse angles (except when you get eight right angles). All the acute angles are congruent, all the obtuse angles are congruent, and every acute angle is supplementary to every obtuse angle.

THEOREMS & POSTULATES

Parallel-lines-with-transversal theorems: If two parallel lines are cut by a transversal, then

>> Corresponding angles are congruent.

>> Alternate interior angles are congruent.

>> Alternate exterior angles are congruent.

>> Same-side interior angles are supplementary.

>> Same-side exterior angles are supplementary.

And now for something completely different — kind of. Say that you don't know that the lines are parallel. Well, all the preceding theorems work in reverse, so you can use the following reverse theorems to prove that lines are parallel.

THEOREMS & POSTULATES

Lines-cut-by-a-transversal theorems: Two lines are parallel if they're cut by a transversal such that

>> Two corresponding angles are congruent.

>> Two alternate interior angles are congruent.

>> Two alternate exterior angles are congruent.

>> Two same-side interior angles are supplementary.

>> Two same-side exterior angles are supplementary.

EXAMPLE

Q. Given: $a \parallel b$

Find: $\angle 1$

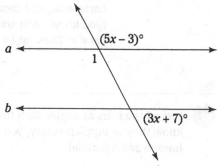

A. The $(5x - 3)°$ angle and the $(3x + 7)°$ angle are same-side exterior angles, so according to the theorem, they're supplementary. Supplementary angles add up to 180°, so

$$(5x - 3) + (3x + 7) = 180$$
$$8x + 4 = 180$$
$$8x = 176$$
$$x = 22$$

Plugging 22 into $(5x - 3)°$ gives you 107° for that angle, and because that angle and $\angle 1$ are vertical angles (see Chapter 2), $\angle 1$ is also 107°.

EXAMPLE

Q. Given: $\angle 1 \cong \angle 2$

$\angle G \cong \angle P$

$\overline{GI} \cong \overline{PU}$

Prove: $\overline{BG} \parallel \overline{WP}$

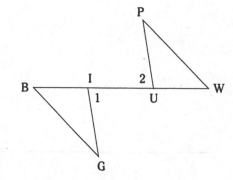

A.

Statements	Reasons
1) $\angle 1 \cong \angle 2$	1) Given.
2) $\angle BIG \cong \angle WUP$	2) Supplements of congruent angles are congruent.
3) $\overline{GI} \cong \overline{PU}$	3) Given.
4) $\angle G \cong \angle P$	4) Given.
5) $\triangle BIG \cong \triangle WUP$	5) ASA (2, 3, 4).
6) $\angle B \cong \angle W$	6) CPCTC.
7) $\overline{BG} \parallel \overline{WP}$	7) If alternate interior angles are congruent, then lines are parallel.

Tip: If you have any difficulty seeing that $\angle B$ and $\angle W$ are indeed alternate interior angles, rotate the figure counterclockwise till the parallel segments $\overline{PW}$ and $\overline{BG}$ are horizontal, and then extend $\overline{PW}$, $\overline{BG}$, and $\overline{BW}$ in both directions, turning them into lines (you know, with arrows). After you do this, you'll be looking at the familiar scheme of parallel lines cut by a transversal, like in Figure 6-1.

---

1 List all the pairs of angles such that if you know they're supplementary, you can prove lines *m* and *n* parallel.

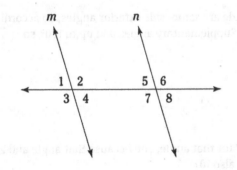

2 Are lines *p* and *q* parallel?

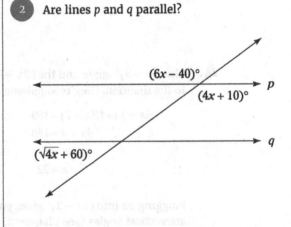

*3 Given that lines *o* and *z* are parallel, solve for *x*. **Hint:** You don't need to use anywhere near all the angles I've numbered. (If I had numbered only the angles you need, I would've given away the solution.)

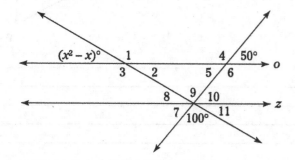

*4 Given: Circle Q

           $\overline{TP} \cong \overline{SR}$

           $m\angle Q = 110$

Prove: $\overline{TS} \parallel \overline{PR}$ (paragraph proof)

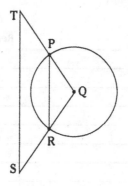

5   Given: $\overline{BL}$ bisects $\angle QBP$

            $\overline{BL} \parallel \overline{JP}$

Prove: $\triangle PBJ$ is isosceles

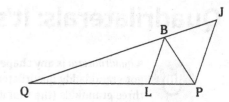

Statements	Reasons

6 Given: $\overline{RA} \parallel \overline{TI}$

$\overline{RA} \cong \overline{TI}$

$\overline{DI} \cong \overline{NA}$

Prove: $\overline{DR} \parallel \overline{TN}$

Statements	Reasons

Quadrilaterals: It's a Family Affair

REMEMBER

A *quadrilateral* **is any shape with four straight sides.** In the family tree of quadrilaterals, you've got granddaddy quadrilateral, his three kids (the kite, the parallelogram, and the trapezoid), three grandkids (the rhombus, the rectangle, and the isosceles trapezoid), and a single great-grandchild (the square). Check out the family tree in Figure 6-2 and the definitions that follow.

>> **Kite:** A quadrilateral in which two disjoint pairs of consecutive sides are congruent (in other words, one side can't be used in both pairs) — it often looks just like the kites you're used to

>> **Parallelogram:** A quadrilateral that has two pairs of parallel sides

>> **Rhombus:** A quadrilateral with four congruent sides; a rhombus is both a kite and a parallelogram

>> **Rectangle:** A quadrilateral with four right angles; a rectangle is a parallelogram

>> **Square:** A quadrilateral with four congruent sides and four right angles; a square is both a rhombus and a rectangle

>> **Trapezoid:** A quadrilateral with exactly one pair of parallel sides; the parallel sides are called the *bases,* and the nonparallel sides are the *legs*

>> **Isosceles trapezoid:** A trapezoid with congruent legs

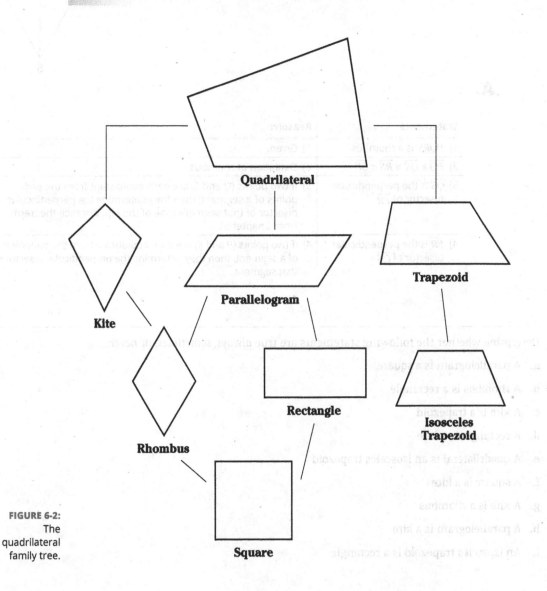

FIGURE 6-2:
The quadrilateral family tree.

Q. Given: Rhombus *PQRS*

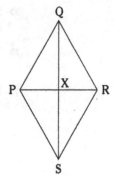

EXAMPLE

Prove: Its diagonals are perpendicular bisectors of each other

A.

Statements	Reasons
1) *PQRS* is a rhombus	1) Given.
2) $\overline{PQ} \cong \overline{QR} \cong \overline{RS} \cong \overline{SP}$	2) Definition of rhombus.
3) $\overline{QS}$ is the perpendicular bisector of $\overline{PR}$	3) If two points (*Q* and *S*) are each equidistant from the end-points of a segment, then they determine the perpendicular bisector of that segment (one of the equidistance theorems from Chapter 5).
4) $\overline{PR}$ is the perpendicular bisector of $\overline{QS}$	4) If two points (*P* and *R*) are each equidistant from the endpoints of a segment, then they determine the perpendicular bisector of that segment.

 Determine whether the following statements are true *always*, *sometimes*, or *never*:

a. A parallelogram is a square

b. A rhombus is a rectangle

c. A kite is a trapezoid

d. A rectangle is a kite

e. A quadrilateral is an isosceles trapezoid

f. A square is a kite

g. A kite is a rhombus

h. A parallelogram is a kite

i. An isosceles trapezoid is a rectangle

8 Given: *JOHN* is a parallelogram

Prove: $\angle J \cong \angle H$

Statements	Reasons

9 Given: *MARY* is a parallelogram

Prove: $\overline{MA} \cong \overline{RY}$

Hint: The figure is incomplete. Use your drawing skills.

Statements	Reasons

Given: *TRAP* is a trapezoid with bases $\overline{TR}$ and $\overline{PA}$

$\triangle TRI$ is isosceles with bases $\overline{TR}$ and $\overline{PA}$

Prove: *TRAP* is isosceles

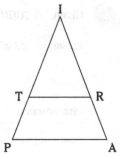

Statements	Reasons

Discovering the Properties of the Parallelogram and the Kite

You may be wondering what these two quadrilaterals have in common and why they're in this section together. I hate to disappoint you, but they have little in common, and I put them together because I couldn't fit all the quadrilaterals in one section and I had to split them up somehow! So, without further ado, let me present to you the properties of the parallelogram and the kite.

REMEMBER

Properties of the parallelogram (see Figure 6-3):

>> Opposite sides are parallel by definition.

>> Opposite sides are congruent.

» Opposite angles are congruent.

» Consecutive angles are supplementary (for example, ∠BCD is supplementary to ∠CDA).

» The diagonals bisect each other.

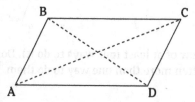

FIGURE 6-3:
ABCD is a parallelogram.

REMEMBER

Properties of the kite (see Figure 6-4):

» Two disjoint pairs of consecutive sides are congruent by definition ($\overline{PQ} \cong \overline{RQ}$ and $\overline{PS} \cong \overline{RS}$).

» The diagonals are perpendicular.

» One diagonal ($\overline{QS}$, the *main diagonal*) is the perpendicular bisector of the other diagonal ($\overline{PR}$, the *cross diagonal*). ("Main diagonal" and "cross diagonal" are good and useful terms, but you won't find them in other geometry books because I made them up.)

» The main diagonal bisects a pair of opposite angles (∠Q and ∠S).

» The opposite angles at the endpoints of the cross diagonal are congruent (∠P and ∠R).

These last three properties are called the *half properties* of the kite.

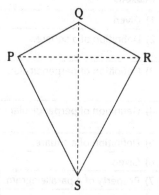

FIGURE 6-4:
PQRS is a kite.

Q. Given: *MINT* is a parallelogram

GILT is a square

Prove: △*MIG* ≅ △*NTL*

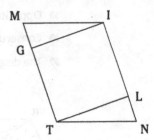

EXAMPLE

A. I do this proof two different ways (I know of at least four ways to do it). Don't forget that when it comes to proofs, there's often more than one way to do them.

Method 1

Statements	Reasons
1) *MINT* is a parallelogram	1) Given.
2) $\overline{MT} \cong \overline{IN}$	2) Property of a parallelogram.
3) *GILT* is a square	3) Given.
4) $\overline{GT} \cong \overline{LI}$	4) Definition of a square.
5) $\overline{MG} \cong \overline{NL}$	5) Segment subtraction.
6) ∠*M* ≅ ∠*N*	6) Property of a parallelogram.
7) $\overline{MI} \cong \overline{NT}$	7) Property of a parallelogram.
8) △*MIG* ≅ △*NTL*	8) SAS (5, 6, 7).

Method 2

Statements	Reasons
1) *GILT* is a square	1) Given.
2) ∠*IGT* is a right angle ∠*TLI* is a right angle	2) Definition of a square.
3) $\overline{IG} \perp \overline{MT}$ $\overline{TL} \perp \overline{NI}$	3) Definition of perpendicular.
4) ∠*MGI* is a right angle ∠*NLT* is a right angle	4) Definition of perpendicular.
5) $\overline{IG} \cong \overline{TL}$	5) Definition of a square.
6) *MINT* is a parallelogram	6) Given.
7) $\overline{MI} \cong \overline{NT}$	7) Property of a parallelogram.
8) △*MIG* ≅ △*NTL*	8) HLR (7, 5, 4).

11 Given: PCTR is a kite

$\overline{PC} \cong \overline{PR}$

Prove: $\angle CAT \cong \angle RAT$

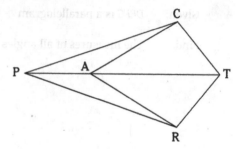

Statements	Reasons

*12 Given: KITE is — guess what — a kite

Find: The lengths of all sides

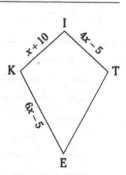

Given: *DEFG* is a parallelogram

Find: The measures of all angles

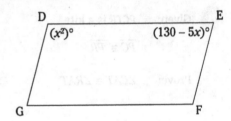

 Given: *YZGH* is a parallelogram

$\overline{ZA} \cong \overline{HO}$

Prove: $\overline{YO} \cong \overline{GA}$

This proof can be done two ways, using two different pairs of congruent triangles.

Hint: With either method, the final reason of the proof is — as I'm sure you can guess — CPCTC.

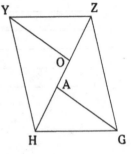

Statements	Reasons

15   Given: *NQRM* is a parallelogram

                $\overline{NO} \cong \overline{RS}$

    Prove: $\overline{SL} \cong \overline{OP}$

As with problem 14, you can do this proof two ways, again using two different pairs of congruent triangles.

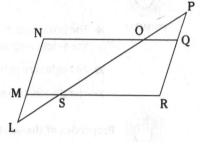

Statements	Reasons

Properties of Rhombuses, Rectangles, and Squares

Keep referring to the quadrilateral family tree in Figure 6-2. Knowing how the different quadrilaterals are related to each other can really help you remember their properties. For example, you can find out from the preceding section that the diagonals of a parallelogram bisect each other. So, because rhombuses, rectangles, and squares are all parallelograms, they automatically share that property.

REMEMBER

Properties of the rhombus (see Figure 6-5):

» The properties of a parallelogram apply (the ones that matter here are parallel sides, opposite angles congruent, and consecutive angles supplementary).

» All sides are congruent by definition.

» The diagonals bisect the angles.

» The diagonals are perpendicular bisectors of each other.

Properties of the rectangle:

» The properties of a parallelogram apply (the ones that matter here are parallel sides, opposite sides congruent, and diagonals bisect each other).

» All angles are right angles by definition.

» The diagonals are congruent.

REMEMBER

Properties of the square:

» The properties of a rhombus apply (the ones that matter here are parallel sides, diagonals are perpendicular bisectors of each other, and diagonals bisect the angles).

» The properties of a rectangle apply (the only one that matters here is diagonals are congruent).

» All sides are congruent by definition.

» All angles are right angles by definition.

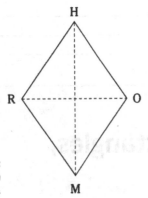

FIGURE 6-5:
RHOM is a rhombus.

EXAMPLE

Q. If a rhombus has sides of length 10 and one diagonal measuring 12, what's the length of the other diagonal?

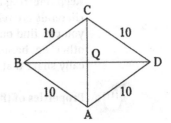

A. The diagonals in a rhombus are perpendicular bisectors of each other, so if *AC* is 12, *AQ* must be 6. $\triangle ABQ$ is a right triangle, so you can solve for *BQ* with the Pythagorean Theorem or by recognizing that $\triangle ABQ$ is in the $3:4:5$ family (see Chapter 4). Either way, *BQ* is 8, and thus, *BD* is 16. But $\overline{BD}$ might be the diagonal that has a length of 12. If that's the case, then *AC* ends up being 16.

PART 3 **Polygons, Proof and Non-Proof Problems**

Q. Given: *BORE* is a rectangle

   $OB = DR = 10$

Find: The length of $\overline{BE}$

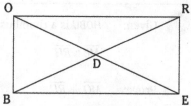

A. In a rectangle, the diagonals are congruent and they bisect each other, so because *DR* is 10, both diagonals are 20. You have right triangle *BOE* with a leg of 10 and a hypotenuse of 20. You should recognize this figure as a 30° – 60° – 90° right triangle (see Chapter 4) where *OB* is *x* and *BE* is $x\sqrt{3}$. $\overline{BE}$ thus has a length of $10\sqrt{3}$.

---

16 Given: *RECT* is a rectangle

   Find: *x* and *y*

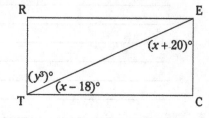

---

17 Given: *MATH* is a rhombus

   ∠*QMA* measures 60°

   HQ is 8

   Find: Measures of all sides and angles in △*QAT*

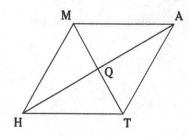

18 Given: *HOBU* is a rhombus

$\overline{MB} \cong \overline{RH}$

Prove: $\overline{MO} \parallel \overline{RU}$

Note: You can do this proof in a few different ways.

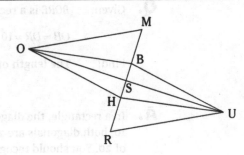

Statements	Reasons

19 Given: *ANGL* is a rectangle

Find: *x, y, z,* and *AL*

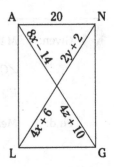

20 Given: *QRVT* is a rhombus

Prove: $\angle 1 \cong \angle 2$

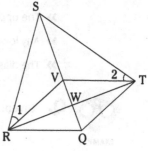

Statements	Reasons

Unearthing the Properties of Trapezoids and Isosceles Trapezoids

I bet you're dying to add to your list of properties, so here you go: Last but not least, the trapezoid and the isosceles trapezoid.

REMEMBER

Properties of the trapezoid:

>> The bases are parallel by definition.

>> Each lower base angle is supplementary to the adjacent upper base angle.

REMEMBER

Properties of the isosceles trapezoid:

>> The properties of a trapezoid apply by definition (parallel bases).

>> The legs are congruent by definition.

» The lower base angles are congruent.

» The upper base angles are congruent.

» Any lower base angle is supplementary to any upper base angle.

» The diagonals are congruent.

Q. Given: PQRS is an isosceles trapezoid with bases $\overline{PS}$ and $\overline{QR}$

$\angle PQX$ is 85°

$\angle PSR$ is 75°

EXAMPLE

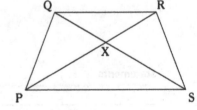

Find: $\angle QXR$

A. An isosceles trapezoid has congruent legs and congruent diagonals. Using those congruent legs and diagonals and $\overline{PS} \cong \overline{PS}$, you get $\triangle PQS = \triangle SRP$ by SSS. $\angle SRX$ is congruent to $\angle PQX$ by CPCTC, so $\angle SRX$ is 85°. The angles in $\triangle PRS$ need to add up to 180°, and you have 85° and 75° ($\angle PSR$) so far, so $\angle RPS$ has to be 20°. $\angle QSP$ is congruent to $\angle RPS$ by CPCTC, so it's 20° as well. The angles in $\triangle PXS$ have to add up to 180°, so $\angle PXS$ is 180 – 20 – 20, or 140°. Finally, $\angle QXR$ and $\angle PXS$ are vertical angles, so $\angle QXR$ is also 140°.

---

 21 Given: QXJW is an isosceles trapezoid with bases $\overline{QW}$ and $\overline{XJ}$

Prove: $\triangle QXZ \cong \triangle WJZ$

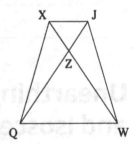

Statements	Reasons

*22 Given: *ZOID* is a trapezoid with bases $\overline{ZD}$ and $\overline{OI}$

                  △*ZCD* is isosceles with base $\overline{ZD}$

Prove: *ZOID* is isosceles

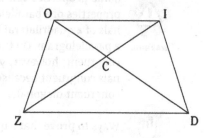

Hint: Before beginning this problem, you may want to review the ten strategies for proofs in Chapter 16. Three of the strategies are especially helpful for this problem.

Statements	Reasons

Proving That a Quadrilateral Is a Parallelogram or a Kite

In the previous few sections, you can find the definitions of the various quadrilaterals and their properties. In this and the next section, "Proving That a Quadrilateral Is a Rhombus, Rectangle, or Square," you move on to proving that a specific quadrilateral is of a particular type. These three things (definitions, properties, and methods of proof) are related, but there are important differences among them.

You can always use a defining characteristic of a particular quadrilateral to prove that a figure is that particular quadrilateral. For example, a parallelogram is *defined* as a quadrilateral with two pairs of parallel sides, and you can *prove* that a quadrilateral is a parallelogram by showing just that. With other properties, however, the process isn't so simple.

WARNING

Some properties can be used as a proof method, but others cannot. For example, one of the properties of a parallelogram is that its diagonals bisect each other, and proving that the diagonals of a quadrilateral bisect each other is one of the five ways of proving that a quadrilateral is a parallelogram. On the other hand, one of the properties of a rectangle is that its diagonals are congruent; however, you *can't* prove that a quadrilateral is a rectangle by showing its diagonals congruent because some kites, isosceles trapezoids, and no-name quadrilaterals also have congruent diagonals.

REMEMBER

Ways to prove that a quadrilateral is a parallelogram: You have a parallelogram if

>> Both pairs of opposite sides of a quadrilateral are parallel (reverse of definition).

>> Both pairs of opposite sides of a quadrilateral are congruent (converse of property).

>> Both pairs of opposite angles of a quadrilateral are congruent (converse of property).

>> The diagonals of a quadrilateral bisect each other (converse of property).

>> One pair of opposite sides of a quadrilateral are both parallel and congruent (note that this is neither the reverse of the definition nor the converse of a property).

REMEMBER

Ways to prove that a quadrilateral is a kite: You've got a kite if

>> Two disjoint pairs of consecutive sides of a quadrilateral are congruent (reverse of definition).

>> One of the diagonals of a quadrilateral is the perpendicular bisector of the other (converse of property).

EXAMPLE

Q. Given: ∠*NOT* ≅ ∠*BAD*

∠*DOA* ≅ ∠*TAO*

Prove: *DOTA* is a parallelogram

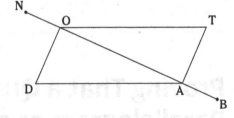

A.

Statements	Reasons
1) ∠*NOT* ≅ ∠*BAD*	1) Given.
2) $\overline{OT}$ ∥ $\overline{DA}$	2) If alternate exterior angles are congruent, then lines are parallel.
3) ∠*DOA* ≅ ∠*TAO*	3) Given.
4) $\overline{DO}$ ∥ $\overline{AT}$	4) If alternate interior angles are congruent, then lines are parallel.
5) *DOTA* is a parallelogram	5) If both pairs of opposite sides of a quadrilateral are parallel, then it is a parallelogram.

Q. Given: ∠1 ≅ ∠2

$\overline{LY}$ bisects ∠GLF

Prove: GOFL is a kite

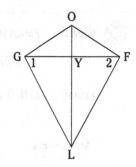

A.

Statements	Reasons
1) ∠1 ≅ ∠2	1) Given.
2) $\overline{GL} \cong \overline{FL}$	2) If angles, then sides.
3) $\overline{LY}$ bisects ∠GLF	3) Given.
4) ∠GLY ≅ ∠FLY	4) Definition of bisect.
5) $\overline{LO} \cong \overline{LO}$	5) Reflexive.
6) ΔGLO ≅ ΔFLO	6) SAS (2, 4, 5).
7) $\overline{GO} \cong \overline{FO}$	7) CPCTC.
8) GOFL is a kite	8) If two disjoint pairs of consecutive sides of a quadrilateral are congruent, then it is a kite (Statements 2 and 7).

---

 23 Given: *DEAL* is a parallelogram

∠*DEN* ≅ ∠*ALO*

Prove: *NEOL* is a parallelogram

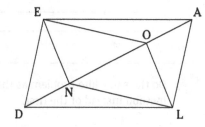

Statements	Reasons

24 Given: *EMNA* is a parallelogram

$\overline{XE} \cong \overline{RN}$ and $\overline{LE} \cong \overline{IN}$

Prove: *LXIR* is a parallelogram

Statements	Reasons

25 Do the same kite problem as the *GOFLY* example problem, but this time use the second kite proof method instead of the first.

Statements	Reasons

26 Given: Diagram as shown

Prove: *JKLM* is a parallelogram (paragraph proof)

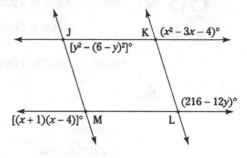

Proving That a Quadrilateral Is a Rhombus, Rectangle, or Square

Note that when proving that a quadrilateral is a rhombus, rectangle, or square, you sometimes go directly from, say, quadrilateral to rhombus or quadrilateral to square — like you do when proving a quadrilateral to be a parallelogram or a kite. But at other times, you first have to prove (or be given) that the quadrilateral is a particular quadrilateral. For example, some methods for proving that a quadrilateral is a rhombus require that you know that the quadrilateral is a parallelogram.

REMEMBER

Ways to prove that a quadrilateral is a rhombus: You have a rhombus if

>> All sides of a quadrilateral are congruent (reverse of definition).

>> The diagonals of a quadrilateral bisect all angles (converse of property).

>> The diagonals of a quadrilateral are perpendicular bisectors of each other (converse of property).

>> Two consecutive sides of a parallelogram are congruent.

>> Either diagonal of a parallelogram bisects two angles.

>> The diagonals of a parallelogram are perpendicular.

REMEMBER

Ways to prove that a quadrilateral is a rectangle: You've got a rectangle if

>> All angles in a quadrilateral are right angles (reverse of definition).

>> A parallelogram contains a right angle.

>> The diagonals of a parallelogram are congruent.

REMEMBER

Ways to prove that a quadrilateral is a square: You have a square if

>> A quadrilateral has four congruent sides and four right angles (reverse of definition).

>> A quadrilateral is both a rhombus and a rectangle.

Q. Given: *KNOT* is a parallelogram.

∠*TOK* ≅ ∠*NOK*

Prove: *KNOT* is a rhombus

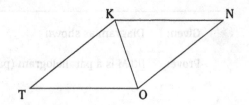

A.

Statements	Reasons
1) ∠*TOK* ≅ ∠*NOK*	1) Given.
2) *KNOT* is a parallelogram	2) Given.
3) $\overline{KN} \parallel \overline{TO}$	3) Property of a parallelogram.
4) ∠*TOK* ≅ ∠*NKO*	4) Alternate interior angles are congruent.
5) ∠*NOK* ≅ ∠*NKO*	5) Transitivity.
6) $\overline{NK} \cong \overline{NO}$	6) If angles, then sides.
7) *KNOT* is a rhombus	7) If one pair of consecutive sides of a parallelogram is congruent, then it is a rhombus.

27) Given: *QVST* is a parallelogram

∠*UQR* ≅ ∠*USR*

Prove: *QVST* is a rhombus

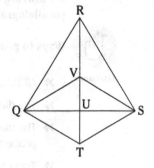

Statements	Reasons

Solutions

(1) For the lines to be parallel, you need the same side interior angles or the same side exterior angles to be supplementary, according the interior- and exterior-angle theorems. Therefore, these theorems give you the following pairs: angles 2 and 5 and angles 4 and 7 (the same-side interior angles) and angles 1 and 6 and angles 3 and 8 (the same-side exterior angles).

Congrats if you saw that more answers are correct. If you recognize, for example, that angles 1 and 7 are supplementary, you can prove *m* parallel to *n*. Why? Angles 6 and 7 are congruent vertical angles, right? So, if angles 1 and 7 are supplementary, angles 1 and 6 have to be supplementary as well. Then the theorem tells you that the lines are parallel. By the same logic, the following supplementary pairs also allow you to prove the lines parallel: angles 2 and 8, angles 3 and 5, and angles 4 and 6.

(2) The two angles on the top are vertical angles; thus, because all vertical angles are congruent,

$$6x - 40 = 4x + 10$$
$$2x = 50$$
$$x = 25$$

Plugging $x = 25$ into $(6x - 40)°$ gives you 110° for that angle.

Plugging $x = 25$ into $(\sqrt{4x} + 60)°$ gives you 70° for that angle.

Because those two angles are same-side exterior angles, they have to be supplementary for the lines to be parallel. 110° and 70° sum to 180° so, yes, the lines are parallel.

(*3) *Warning:* When working on a problem that has more than one transversal, make sure you *use only one transversal at a time* when using the theorems to compare various angles. For example, *the theorems tell you nothing* about how angles 4 and 11 compare to each other because those angles use two different transversals. Ditto for angles 2 and 10.

Here goes. The 50° angle and ∠10 are corresponding angles, so ∠10 is also 50°. Next, angles 10 and 11 have to add up to 80°, because together they form an angle that's supplementary to the 100° angle. That makes ∠11 = 30°. Because the $x^2 - x$ angle and ∠11 are alternate exterior angles, they're congruent, so $x^2 - x = 30$. Now you can solve for x:

$$x^2 - x = 30$$
$$x^2 - x - 30 = 0$$
$$(x - 6)(x + 5) = 30$$
$$x = 6 \text{ or } -5$$

Note that both 6 and −5 are valid answers. In geometry problems, you often reject negative answers because segment lengths and angle measures can't be negative. But here, plugging −5 into $(x^2 - x)°$ gives you a *positive* angle (namely 30°), so −5 is a perfectly good answer.

(*4) The two radii are congruent, so △PQR is isosceles (see Chapter 4). That makes ∠QPR congruent to ∠QRP. Then, to make the angles in △PQR add up to 180°, ∠QPR and ∠QRP must each be 35°.

Because $\overline{TP}$ and $\overline{SR}$ are congruent, you can add them to the congruent radii, making $\overline{TQ}$ and $\overline{SQ}$ congruent. Thus, the big triangle is isosceles with $\angle T \cong \angle S$. By the same reasoning as before, $\angle T$ and $\angle S$ are also 35° angles. Finally, $\angle T$ and $\angle QPR$ are corresponding angles. Because they're both 35°, the lines have to be parallel.

(5)

Statements	Reasons
1) $\overline{BL}$ bisects $\angle QBP$	1) Given.
2) $\angle QBL \cong \angle LBP$	2) Definition of bisect.
3) $\overline{BL} \parallel \overline{JP}$	3) Given.
4) $\angle QBL \cong \angle J$	4) If lines are parallel, then corresponding angles are congruent (using transversal $\overline{QJ}$).
5) $\angle LBP \cong \angle BPJ$	5) If lines are parallel, then alternate interior angles are congruent (using transversal $\overline{BP}$).
6) $\angle J \cong \angle BPJ$	6) Transitivity (2, 4, 5).
7) $\overline{BP} \cong \overline{BJ}$	7) If angles, then sides.
8) $\triangle PBJ$ is isosceles	8) Definition of isosceles triangle.

(6)

Statements	Reasons
1) $\overline{RA} \cong \overline{TI}$	1) Given.
2) $\overline{RA} \parallel \overline{TI}$	2) Given.
3) $\angle DAR \cong \angle NIT$	3) If lines are parallel, then alternate exterior angles are congruent.
4) $\overline{DI} \cong \overline{NA}$	4) Given.
5) $\overline{DA} \cong \overline{NI}$	5) Segment subtraction (subtracting $\overline{AI}$ from both segments).
6) $\triangle DAR \cong \triangle NIT$	6) SAS (1, 3, 5).
7) $\angle D \cong \angle N$	7) CPCTC.
8) $\overline{DR} \parallel \overline{TN}$	8) If alternate interior angles are congruent, then the lines are parallel.

(7) One way to do these always–sometimes–never problems is to look at the quadrilateral family tree and follow these guidelines:

TIP

- If you go up from the first figure to the second, the answer is *always*.

- If you go down, the answer is *sometimes*.

- If you can make the connection by going down and then up (like from a kite to a rectangle or vice versa), it's *sometimes*.

- And if the only way to make the connection is by going up and then down, the answer is *never*.

Here are the correct answers to problem 7:

a. Sometimes

b. Sometimes (when it's a square)

c. Never

d. Sometimes (when it's a square)

e. Sometimes

f. Always

g. Sometimes

h. Sometimes (when it's a rhombus)

i. Never

(8)

Statements	Reasons
1) *JOHN* is a parallelogram	1) Given.
2) $\overline{JO} \cong \overline{NH}$	2) Definition of parallelogram.
3) ∠J is supplementary to ∠N	3) If lines are parallel, then same-side interior angles are supplementary (using transversal $\overline{JN}$).
4) $\overline{JN} \cong \overline{OH}$	4) Definition of parallelogram.
5) ∠H is supplementary to ∠N	5) If lines are parallel, then same-side interior angles are supplementary (using transversal $\overline{NH}$).
6) ∠J ≅ ∠H	6) Supplements of the same angle are congruent.

(9)

Statements	Reasons
1) *MARY* is a parallelogram	1) Given.
2) Draw $\overline{AY}$ ($\overline{MR}$ would work as well)	2) Two points determine a line.
3) $\overline{MA} \parallel \overline{YR}$	3) Definition of parallelogram.
4) ∠MAY ≅ ∠RYA	4) If lines are parallel, then alternate interior angles are congruent (using parallel segments $\overline{MA}$ and $\overline{YR}$ and transversal $\overline{AY}$).
5) $\overline{AR} \parallel \overline{MY}$	5) Definition of parallelogram.
6) ∠MYA ≅ ∠RAY	6) If lines are parallel, then alternate interior angles are congruent (using parallel segments $\overline{AR}$ and $\overline{MY}$ with transversal $\overline{AY}$).
7) $\overline{AY} \cong \overline{YA}$	7) Reflexive.
8) △MAY ≅ △RYA	8) ASA (4, 7, 6).
9) $\overline{MA} \cong \overline{RY}$	9) CPCTC.

10	Statements	Reasons
	1) $\triangle TRI$ is isosceles with base $\overline{TR}$	1) Given.
	2) $\overline{TI} \cong \overline{RI}$	2) Definition of isosceles triangle.
	3) $\angle ITR \cong \angle IRT$	3) If sides, then angles.
	4) $TRAP$ is a trapezoid with bases $\overline{TR}$ and $\overline{PA}$	4) Given.
	5) $\overline{TR} \parallel \overline{PA}$	5) Definition of trapezoid.
	6) $\angle P \cong \angle ITR$	6) If lines are parallel, then corresponding angles are congruent (using transversal $\overline{IP}$).
	7) $\angle A \cong \angle IRT$	7) If lines are parallel, then corresponding angles are congruent (using transversal $\overline{IA}$).
	8) $\angle P \cong \angle A$	8) Transitivity (3, 6, 7).
	9) $\overline{IP} \cong \overline{IA}$	9) If angles, then sides.
	10) $\overline{PT} \cong \overline{AR}$	10) Subtraction (Statements 2 and 9).
	11) $TRAP$ is isosceles	11) Definition of isosceles trapezoid.

11	Statements	Reasons
	1) $PCTR$ is a kite	1) Given.
	2) $\overline{PC} \cong \overline{PR}$	2) Given.
	3) $\overline{TC} \cong \overline{TR}$	3) Property of a kite (because $\overline{PC} \cong \overline{PR}$, the other disjoint pair of sides must also be congruent).
	4) $\overline{TP}$ bisects $\angle CTR$	4) Property of a kite.
	5) $\angle CTP \cong \angle RTP$	5) Definition of bisect.
	6) $\overline{AT} \cong \overline{AT}$	6) Reflexive.
	7) $\triangle CAT \cong \triangle RAT$	7) SAS (3, 5, 6).
	8) $\angle CAT \cong \angle RAT$	8) CPCTC.

*12) Set the consecutive sides $\overline{KI}$ and $\overline{TI}$ equal to each other (because, by definition, two pairs of consecutive sides are congruent), and solve:

$$x + 10 = 4x - 5$$
$$-3x = -15$$
$$x = 5$$

Plugging $x = 5$ into the three sides gives you $5 + 10$, or 15 for KI; $4(5) - 5$, or 15 for TI; and $6(5) - 5$, or 25 for KE. $\overline{TE}$ must be congruent to $\overline{KE}$, so it also measures 25.

But wait! There's another possibility. Don't forget — figures don't have to be drawn to scale. $\overline{KI}$ and $\overline{KE}$ could be a congruent pair of sides with $\overline{TI}$ and $\overline{TE}$ the other congruent pair. Thus,

$$x + 10 = 6x - 5$$
$$-5x = -15$$
$$x = 3$$

Doing the math with $x = 3$ gives you the following equally valid set of lengths:

$$KI = KE = 13$$
$$TI = TE = 7$$

(13) Consecutive angles in a parallelogram are supplementary, so

$$x^2 + (130 - 5x) = 180$$

This is a quadratic equation, so set it equal to zero and solve by factoring (you can also use the quadratic formula, of course).

$$x^2 - 5x - 50 = 0$$
$$(x - 10)(x + 5) = 0$$
$$x = 10 \ \text{ or } \ -5$$

Plugging $x = 10$ into angles D and E gives you $100°$ for $\angle D$ and $80°$ for $\angle E$. $\angle G$ must also be $80°$, and $\angle F$ is $100°$.

But don't reject $x = -5$ just because it's a negative number. Plugging $x = -5$ into angles D and E gives you another valid set of angles (albeit with measures *way* different from what the angles look like in the figure): $\angle D = \angle F = 25°$ and $\angle E = \angle G = 155°$.

(14) **Method 1**

Statements	Reasons
1) $YZGH$ is a parallelogram	1) Given.
2) $\overline{ZG} \cong \overline{YH}$	2) Property of a parallelogram.
3) $\overline{ZG} \parallel \overline{YH}$	3) Property of a parallelogram.
4) $\angle GZA \cong \angle YHO$	4) Alternate interior angles are congruent (using transversal $\overline{ZH}$).
5) $\overline{ZA} \cong \overline{HO}$	5) Given.
6) $\triangle GZA \cong \triangle YHO$	6) SAS (2, 4, 5).
7) $\overline{YO} \cong \overline{GA}$	7) CPCTC.

Method 2

Statements	Reasons
1) *YZGH* is a parallelogram	1) Given.
2) $\overline{YZ} \cong \overline{HG}$	2) Property of a parallelogram.
3) $\overline{YZ} \parallel \overline{HG}$	3) Property of a parallelogram.
4) $\angle YZO \cong \angle GHA$	4) Alternate interior angles are congruent (using transversal $\overline{ZH}$).
5) $\overline{ZA} \cong \overline{HO}$	5) Given.
6) $\overline{ZO} \cong \overline{HA}$	6) Subtraction.
7) $\triangle YZO \cong \triangle GHA$	7) SAS (2, 4, 6).
8) $\overline{YO} \cong \overline{GA}$	8) CPCTC.

(15) **Method 1 (Using $\triangle NLO$ and $\triangle RPS$)**

Statements	Reasons
1) *NQRM* is a parallelogram	1) Given.
2) $\overline{NQ} \parallel \overline{MR}$	2) Property of a parallelogram.
3) $\angle SON \cong \angle OSR$	3) Alternate interior angles are congruent.
4) $\overline{NO} \cong \overline{RS}$	4) Given.
5) $\angle N \cong \angle R$	5) Property of a parallelogram.
6) $\triangle NLO \cong \triangle RPS$	6) ASA (3, 4, 5).
7) $\overline{LO} \cong \overline{PS}$	7) CPCTC.
8) $\overline{SL} \cong \overline{OP}$	8) Subtraction (subtracting $\overline{OS}$ from $\overline{LO}$ and $\overline{PS}$).

Method 2 (Using $\triangle LMS$ and $\triangle PQO$)

Statements	Reasons
1) *NQRM* is a parallelogram	1) Given.
2) $\overline{NM} \parallel \overline{QR}$	2) Property of a parallelogram.
3) $\angle L \cong \angle P$	3) Alternate interior angles are congruent.
4) $\overline{NQ} \parallel \overline{MR}$	4) Property of a parallelogram.
5) $\angle LSM \cong \angle POQ$	5) Alternate exterior angles are congruent.
6) $\overline{NO} \cong \overline{RS}$	6) Given.
7) $\overline{NQ} \cong \overline{MR}$	7) Property of a parallelogram.
8) $\overline{QO} \cong \overline{MS}$	8) Subtraction.
9) $\triangle LSM \cong \triangle POQ$	9) AAS (3, 5, 8).
10) $\overline{SL} \cong \overline{OP}$	10) CPCTC.

16 You have a rectangle, so $\triangle TEC$ is a right triangle. Its angles have to add up to 180°, so because $\angle C$ is a right angle, the other two angles add up to 90°. Thus,

$$(x-18)+(x+20)=90$$
$$2x+2=90$$
$$2x=88$$
$$x=44$$

$\angle ETC$ is thus $44-18$, or 26, and because $\angle RTC$ equals 90°, that leaves 64° for $\angle RTE$:

$$y^3=64$$
$$y=4$$

17 The diagonals in a rhombus are perpendicular bisectors of each other, so because HQ is 8, AQ is also 8. The sides of a rhombus are equal, so $\triangle MAT$ is isosceles with base $\overline{MT}$. The base angles are congruent, so $\angle QTA$, like $\angle QMA$, is 60°. Because $\angle AQT$ is 90°, $\triangle QAT$ is a $30°-60°-90°$ triangle. Its long leg, $\overline{AQ}$, measures 8, so its short leg, $\overline{TQ}$, is $\dfrac{8}{\sqrt{3}}$ (or $\dfrac{8\sqrt{3}}{3}$) units long. The hypotenuse, $\overline{AT}$, is twice that, or $\dfrac{16\sqrt{3}}{3}$ units long. (See Chapter 4 for info on $30°-60°-90°$ triangles.)

18

Statements	Reasons
1) $HOBU$ is a rhombus	1) Given.
2) $\overline{BO} \cong \overline{HU}$	2) Property of a rhombus.
3) $\overline{BO} \parallel \overline{HU}$	3) Property of a rhombus.
4) $\angle OBM \cong \angle UHR$	4) Alternate exterior angles are congruent.
5) $\overline{MB} \cong \overline{RH}$	5) Given.
6) $\triangle OBM \cong \triangle UHR$	6) SAS (2, 4, 5).
7) $\angle M \cong \angle R$	7) CPCTC.
8) $\overline{MO} \parallel \overline{RU}$	8) If alternate interior angles are congruent, then lines are parallel.

19 The diagonals in a rectangle are congruent, and they bisect each other, so all four half-diagonals are equal. You need an equation with a single variable, namely

$$8x-14=4x+6$$
$$4x=20$$
$$x=5$$

Plugging that into $8x-14$ (or $4x+6$) gives you 26 for the length of each half-diagonal. Thus,

$$2y+2=26$$
$$2y=24 \quad \text{and} \quad$$
$$y=12$$

$$4z+10=26$$
$$4z=16$$
$$z=4$$

Finally, the length of diagonal $\overline{LN}$ is twice 26, or 52. You can then compute AL with the Pythagorean Theorem, or — if you're on your toes — you recognize that $\triangle ANL$ is in the $5:12:13$ family (see Chapter 4). AN is 20 (which is $4 \cdot 5$) and LN is 52 (which is $4 \cdot 13$), so AL is 4 times 12, or 48.

20

Statements	Reasons
1) *QRVT* is a rhombus	1) Given.
2) $\overline{VQ}$ is the perpendicular bisector of $\overline{RT}$	2) Property of a rhombus.
3) $\overline{SR} \cong \overline{ST}$	3) If a point is on the perpendicular bisector of a segment, then it is equidistant from the endpoints of that segment (equidistance theorem).
4) $\overline{RV} \cong \overline{TV}$	4) Property of a rhombus.
5) $\overline{SV} \cong \overline{SV}$	5) Reflexive.
6) $\triangle RSV \cong \triangle TSV$	6) SSS (3, 4, 5).
7) $\angle 1 \cong \angle 2$	7) CPCTC.

21

Statements	Reasons
1) *QXJW* is an isosceles trapezoid with bases $\overline{QW}$ and $\overline{XJ}$	1) Given.
2) $\overline{QX} \cong \overline{WJ}$	2) The legs of an isosceles trapezoid are congruent.
3) $\overline{WX} \cong \overline{QJ}$	3) The diagonals of an isosceles trapezoid are congruent.
4) $\overline{QW} \cong \overline{WQ}$	4) Reflexive.
5) $\triangle QXW \cong \triangle WJQ$	5) SSS (2, 3, 4).
6) $\angle QXW \cong \angle WJQ$	6) CPCTC.
7) $\angle QZX \cong \angle WZJ$	7) Vertical angles are congruent.
8) $\triangle QXZ \cong \triangle WJZ$	8) AAS (7, 6, 2).

22

Statements	Reasons
1) $\triangle ZCD$ is isosceles with base $\overline{ZD}$	1) Given.
2) $\overline{ZC} \cong \overline{DC}$	2) Definition of isosceles triangle.
3) $\angle CZD \cong \angle CDZ$	3) If sides, then angles.
4) *ZOID* is a trapezoid with bases $\overline{ZD}$ and $\overline{OI}$	4) Given.
5) $\overline{OI} \parallel \overline{ZD}$	5) The bases of a trapezoid are parallel.
6) $\angle CZD \cong \angle OIZ$	6) Alternate interior angles are congruent (using transversal $\overline{ZI}$).
7) $\angle CDZ \cong \angle IOD$	7) Alternate interior angles are congruent (using transversal $\overline{DO}$).
8) $\angle OIZ \cong \angle IOD$	8) Transitivity (6, 3, 7).
9) $\overline{OC} \cong \overline{IC}$	9) If angles, then sides.
10) $\angle OCZ \cong \angle ICD$	10) Vertical angles are congruent.
11) $\triangle OCZ \cong \triangle ICD$	11) SAS (2, 10, 9).
12) $\overline{OZ} \cong \overline{ID}$	12) CPCTC.
13) *ZOID* is isosceles	13) A trapezoid with congruent legs is isosceles.

(23)

Statements	Reasons
1) *DEAL* is a parallelogram	1) Given.
2) $\overline{DE} \parallel \overline{AL}$	2) Property of parallelogram.
3) $\angle EDN \cong \angle LAO$	3) If lines are parallel, then alternate interior angles are congruent.
4) $\overline{DE} \cong \overline{AL}$	4) Property of parallelogram.
5) $\angle DEN \cong \angle ALO$	5) Given.
6) $\triangle DEN \cong \triangle ALO$	6) ASA (3, 4, 5).
7) $\angle DNE \cong \angle AOL$	7) CPCTC.
8) $\overline{EN} \parallel \overline{OL}$	8) If alternate exterior angles are congruent, then lines are parallel.
9) $\overline{EN} \cong \overline{LO}$	9) CPCTC.
10) *NEOL* is a parallelogram	10) If one pair of opposite sides of a quadrilateral are both parallel and congruent, then it is a parallelogram.

(24)

Statements	Reasons
1) *EMNA* is a parallelogram	1) Given.
2) $\angle E \cong \angle N$	2) Opposite angles of a parallelogram are congruent.
3) $\overline{XE} \cong \overline{RN}$ $\overline{LE} \cong \overline{IN}$	3) Given.
4) $\triangle LEX \cong \triangle INR$	4) SAS (3, 2, 3).
5) $\overline{XL} \cong \overline{RI}$	5) CPCTC.
6) $\overline{EM} \cong \overline{AN}$	6) Opposite sides of a parallelogram are congruent.
7) $\overline{XM} \cong \overline{RA}$	7) Subtraction.
8) $\overline{EA} \cong \overline{MN}$	8) Opposite sides of a parallelogram are congruent.
9) $\overline{LA} \cong \overline{IM}$	9) Subtraction.
10) $\angle A \cong \angle M$	10) Opposite angles of a parallelogram are congruent.
11) $\triangle LAR \cong \triangle IMX$	11) SAS (7, 10, 9).
12) $\overline{LR} \cong \overline{IX}$	12) CPCTC.
13) *LXIR* is a parallelogram	13) If both pairs of opposite sides of a quadrilateral are congruent, then it is a parallelogram.

(25) The first four steps of this proof are the same as in the example problem, so I pick up with step 5.

Statements	Reasons
5) $\triangle GLY \cong \triangle FLY$	5) ASA (1, 2, 4).
6) $\overline{GY} \cong \overline{FY}$	6) CPCTC.
7) $\overline{OL}$ is the perpendicular bisector of $\overline{GF}$	7) If two points are each equidistant from the endpoints of a segment, then they determine the perpendicular bisector of that segment.
8) *GOFL* is a kite	8) If one of the diagonals of a quadrilateral is the perpendicular bisector of the other, then the quadrilateral is a kite.

(*26) I hope this odd problem didn't give you an algebra panic attack. It's not as bad as it looks. The first thing you have to realize is that you don't have to solve for x or y. In fact, solving for the variables is impossible because you don't have any information about the figure that would allow you to write any equations.

First, multiply $(x+1)(x-4)$; that's $x^2 - 3x - 4$. This measure is the same as that of the other x angle, so regardless of the value of x, those angles are congruent. Then, using congruent vertical angles, you can show that $\angle JML$ is congruent to $\angle LKJ$. Now for the y angles. First, simplify $y^2 - (6-y)^2$. That's $y^2 - (36 - 12y + y^2) = 12y - 36$. Then, because $\angle MLK$ is supplementary to the $(216 - 12y)°$ angle, the measure of $\angle MLK = 180 - (216 - 12y) = 12y - 36$. Thus, $\angle MLK$ is congruent to $\angle KJM$. You have two pairs of congruent opposite angles, and so you have a parallelogram.

(27)

Statements	Reasons
1) $\angle UQR \cong \angle USR$	1) Given.
2) $\overline{QR} \cong \overline{SR}$	2) If angles, then sides.
3) $QVST$ is a parallelogram	3) Given.
4) $\overline{VT}$ bisects $\overline{QS}$	4) Property of a parallelogram.
5) $\overline{QU} \cong \overline{SU}$	5) Definition of bisect.
6) $\overline{VT}$ is the perpendicular bisector of $\overline{QS}$	6) If two points are each equidistant from the endpoints of a segment, then they determine the perpendicular bisector of that segment.
7) $QVST$ is a rhombus	7) If the diagonals of a parallelogram are perpendicular, then it is a rhombus.

Chapter 7

Area, Angles, and the Many Sides of Polygon Geometry (No Proofs)

I f you're all proofed-out, you may enjoy this proof-free chapter. Here you work on problems involving formulas for the area of various polygons, the sum of the interior and exterior angles of a polygon, and the number of diagonals of a polygon. If you've always wondered about how many diagonals an octakaidecagon has, you've come to the right place.

Square Units: Finding the Area of Quadrilaterals

You might want to look back at the family tree of quadrilaterals in Chapter 6 — assuming you don't know it by heart — to remind yourself about which quadrilaterals are special cases of other quadrilaterals. Doing so can help you with area problems because when you know, for example, that a rhombus is a special case of both a parallelogram and a kite, you know that you can use either the parallelogram area formula or the kite area formula when computing the area of a rhombus.

Without further ado, here are the area formulas for quadrilaterals.

Quadrilateral area formulas:

» $\text{Area}_{\text{Parallelogram}} = \text{base} \cdot \text{height}$

» $\text{Area}_{\text{Kite}} = \frac{1}{2}\,\text{diagonal}_1 \cdot \text{diagonal}_2$

» $\text{Area}_{\text{Square}} = \text{side}^2$, or $\frac{1}{2}\,\text{diagonal}^2$

» $\text{Area}_{\text{Trapezoid}} = \dfrac{\text{base}_1 + \text{base}_2}{2} \cdot \text{height}$
$= \text{median} \cdot \text{height}$

(The *median* of a trapezoid is the segment that connects the midpoints of the legs. Its length equals the average of the lengths of the bases.)

Ready for more info about quadrilateral area formulas? Here's a handy guide for the quadrilaterals that don't have an area formula in the preceding list:

» For the area of a rectangle, use the parallelogram formula.

» For the area of a rhombus, use either the parallelogram or the kite formula.

» For the area of an isosceles trapezoid, use, of course, the trapezoid formula.

Q. What's the area of parallelogram *ABCD*?

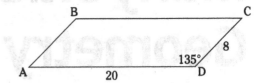

A. *Tip:* For this and many area problems, drawing in altitudes and other perpendicular segments on the diagram can be helpful. And — what often amounts to the same thing — it's a good idea to cut up the figure into right triangles and rectangles.

Draw the altitude from *B* to $\overline{AD}$, and call the length of that segment *h*. ∠*A* is supplementary to ∠*D* (a property of parallelograms; see Chapter 6), so ∠*A* is 45°. Thus, the altitude you drew creates a 45° – 45° – 90° triangle. The hypotenuse, $\overline{AB}$, is congruent to the opposite side, $\overline{CD}$, and therefore has a length of 8. Using 45° – 45° – 90° triangle math, *h* equals $\frac{8}{\sqrt{2}}$ or $4\sqrt{2}$ (see Chapter 4 for details). The area, which equals *base · height*, is thus $20 \cdot 4\sqrt{2}$, or $80\sqrt{2}$ units^2.

Q. Given: Trapezoid *ABCD* with a base of 6, median $\overline{PQ}$ that's 10 units long, and *QD* = 5

Find: Area of *ABCD*

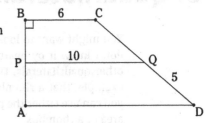

A. For this problem, you can use the trapezoid area formula that uses the median. You know the length of the median, so all you need to compute the area is the trapezoid's height. To get that, first recall that the length of the median, $\overline{PQ}$, is the average of the lengths of the bases, so its length is halfway between them. Since BC is 6, or $10 - 4$, AD must be $10 + 4$, or 14. Next, draw an altitude from C to $\overline{AD}$, creating a right triangle. The length of its hypotenuse, $\overline{CD}$, is twice 5, or 10, and the triangle's base is $14 - 6$, or 8. You have a right triangle in the $3 : 4 : 5$ family (namely a $6 - 8 - 10$ triangle; see Chapter 4), so the altitude is 6 (this has nothing to do, by the way, with the length of $\overline{BC}$, which is coincidentally also 6). Finally, the area, which equals *median · height*, is $10 \cdot 6$, or 60 units^2.

---

1 Given: Parallelogram *PQRS* with sides of 7 and 10 and altitudes h_1 and h_2

Find: The ratio $h_1 : h_2$

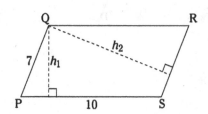

2 Given: Parallelogram *GRAM* as shown

Find: *GRAM*'s area and height

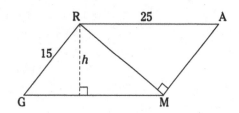

3 Given: Trapezoid *WXYZ* with a perimeter of 35 and an area of 55

Find: *h*

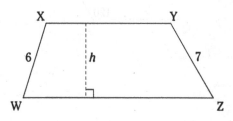

*4 The equation of this circle of radius 8 is $x^2 + y^2 = 64$. Estimate its area using the six trapezoids and two triangles.

Hint: Just calculate the areas of the three trapezoids and one triangle in quadrant I; then multiply your result by 4.

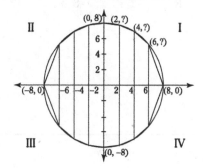

*5 Given: Trapezoid *JKLM* with bases 10 and 18
            and base angles of 60° and 30°

    Find: Area of *JKLM*

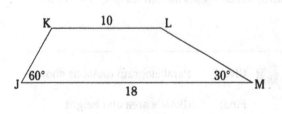

6   Given: Kite *ABCD* as shown, where △*ABC* is
            equilateral

    Find: Area of *ABCD*

    Hint: Use your drawing skills.

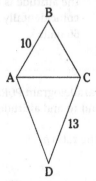

7   Find the area of rhombus *RBUS*

    Hint: Just connect the dots.

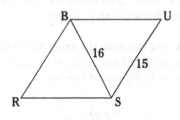

8   Find the area of rhombus *QRST*

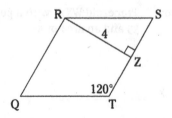

The Standard Formula for the Area of Regular Polygons

Time to cut to the chase. Here's the formula for the area of a regular polygon — but first, its definition: A *regular polygon* is a polygon that's both equilateral (with equal sides) and equiangular (equal angles).

REMEMBER

Area of a regular polygon:

$$\text{Area}_{\text{Reg. Poly.}} = \frac{1}{2}\,\text{perimeter} \cdot \text{apothem}$$

An *apothem* of a regular polygon is a segment joining the polygon's center to the midpoint of any side. It's perpendicular to the side.

This area formula, $A = \frac{1}{2}pa$, is usually written $A = \frac{1}{2}ap$. These formulas are equivalent, of course, so you can use either one, but the way I've written it helps you to think about what you're actually doing. This polygon formula is based on the formula for the area of a triangle, $A = \frac{1}{2}bh$, because when you find the area of a regular polygon, you're essentially dividing the polygon into congruent triangles and finding their areas. Since a polygon's perimeter is the counterpart of the triangles' bases and a polygon's apothem is the counterpart of the triangles' heights, $A = \frac{1}{2}pa$ is the logical way to write the formula.

TIP

Cutting up polygons can be a big help. As you can see in the following example problem, a regular hexagon can be cut into six equilateral triangles, and an equilateral triangle can be cut into two $30° - 60° - 90°$ triangles. For many area problems involving either a hexagon or an equilateral triangle (or both), it's often useful to cut the figure up and make use of one or more $30° - 60° - 90°$ triangles. If, instead, the problem involves a square or a regular octagon, adding the right segments to the diagram produces one or more $45° - 45° - 90°$ triangles that may be the key to the solution. In other polygon problems, cutting up the polygon into some combination of rectangles and these special triangles can help.

An equilateral triangle is a regular polygon, so to figure its area, you can use the regular polygon formula; however, it also has its own area formula. To wit —

REMEMBER

Area of an equilateral triangle:

$$\text{Area}_{\text{Equil. }\triangle} = \frac{s^2\sqrt{3}}{4}$$

EXAMPLE

Q. What's the area of this regular hexagon with a radius of 8? (Yep — that thing is called the *radius*.)

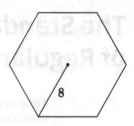

8

A. You can do this problem two ways, using both of the preceding area formulas.

First, draw in the other five radii, and you can see six congruent isosceles triangles.

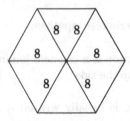

8 8
8   8
8   8

The six angles at the center of the hexagon have to be 60° angles, because all the way around the center is 360°, and 360° ÷ 6 is 60°. An isosceles triangle with a 60° vertex angle is an equilateral triangle, so you have six congruent equilateral triangles.

Method I: Now you can finish using the equilateral triangle formula:

$$\text{Area}_{\text{Equil. }\Delta} = \frac{s^2\sqrt{3}}{4}$$

$$= \frac{8^2\sqrt{3}}{4}$$

$$= 16\sqrt{3}$$

You have six triangles, so the area of the hexagon is $6 \cdot 16\sqrt{3}$, or $96\sqrt{3}$ units^2.

Method II: Draw in the apothem from the center of the hexagon straight down to the midpoint of the bottom side. That apothem cuts the equilateral triangle into two $30° - 60° - 90°$ triangles. These triangles have sides with ratios of $1 : \sqrt{3} : 2$ (see Chapter 4 for more on special right triangles). Each triangle has a hypotenuse of 8, and therefore, a short leg of 4 and a long leg (the apothem) of $4\sqrt{3}$. The perimeter is 6 times 8, or 48, so you're all set to use the polygon formula:

$$\text{Area}_{\text{Reg. Poly.}} = \frac{1}{2}pa$$

$$= \frac{1}{2} \cdot 48 \cdot 4\sqrt{3}$$

$$= 96\sqrt{3} \text{ units}^2$$

9 The span of this regular hexagon is 32. Find its area.

32

 10 Find the area of a regular octagon with sides of length 10.

Hint: Cut up the octagon till you create one or more useful 45° – 45° – 90° triangles.

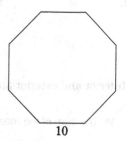

10

More Fantastically Fun Polygon Formulas

The formulas in the preceding section are all about the area of polygons. They're basically meant to show you how much space a polygon takes up. The formulas in this section dive deeper into the building blocks of polygons: the angles and diagonals that give different polygons their unique characteristics.

REMEMBER

Definitions of interior and exterior angles:

>> An *interior* angle of a polygon is an angle inside the polygon at one of its vertices.

>> An *exterior* angle of a polygon is an angle outside the polygon formed by one of its sides and the extension of an adjacent side (see Figure 7-1).

Would you believe me if I told you that regardless of whether a polygon has three sides or a million, the exterior angles of that polygon always add up to 360°? You'd better, because it's a fact. And that's just the beginning of the polygon's special properties.

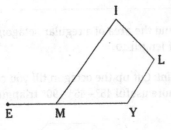

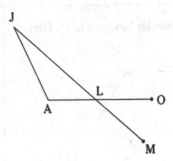

FIGURE 7-1:
∠*EMI* is an exterior angle of quadrilateral *MILY*. Vertical angle ∠*OLM* is *not* an exterior angle of △*AJL*.

REMEMBER

Interior and exterior angle formulas:

>> The sum of the measures of the *interior angles* of a polygon with n sides is $(n-2)180$.

>> The measure of each *interior angle* of an equiangular n-gon is $\dfrac{(n-2)180}{n}$ or $180 - \dfrac{360}{n}$.

>> If you count one exterior angle at each vertex, the sum of the measures of the *exterior angles* of a polygon is always 360°.

>> The measure of each *exterior angle* of an equiangular n-gon is $\dfrac{360}{n}$.

Number of diagonals in a polygon: The number of diagonals that you can draw in an n-gon is $\dfrac{n(n-3)}{2}$.

REMEMBER

EXAMPLE

Q. What's the measure of one of the interior angles of a regular 22-gon, and how many diagonals does it have?

A. A regular polygon is equiangular, so you can use either equiangular formula (in the second bullet). The second version is probably easier to use unless you happen to already know the sum of the interior angles (which is the numerator in the first formula). Note that using the second version amounts to finding the supplement of one of the polygon's exterior angles.

$$\text{Interior angle} = 180 - \frac{360}{n}$$
$$= 180 - \frac{360}{22}$$
$$= 180 - 16\frac{4}{11}$$
$$= 163\frac{7}{11}°$$

$$\text{Diagonals} = \frac{n(n-3)}{2}$$
$$= \frac{22(22-3)}{2}$$
$$= 209$$

11 Given: Hexagon *TAYLOR* with angles as shown

$$\angle ATR \cong \angle O$$

Find: $\angle 1$

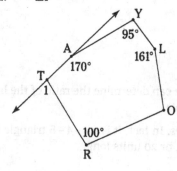

12 Find the number of sides in a polygon whose interior angles add up to

a. 1080°

b. 7920°

c. $(180x^2 + 180)°$ (for some whole number *x*)

d. 825°

13 Find the sum of all exterior angles and, if you have enough information, the measure of one exterior angle in

a. An equiangular pentagon

b. A regular 18-gon

c. An icosagon (20 sides)

d. An equilateral 80-gon

14 How many diagonals can be drawn in a triacontagon (30 sides)?

15 How many sides does a polygon have if it has 3 times as many diagonals as sides?

16 What's the measure of one of the interior angles of an equiangular polygon with 54 diagonals?

Solutions

1 $\overline{RS}$ is congruent to $\overline{PQ}$ (property of a parallelogram; see Chapter 6), so RS is also 7. The area of a parallelogram equals *base · height*, and, obviously, the area must come out the same regardless of which base you use, $\overline{PS}$ or $\overline{RS}$. Thus,

$$\text{base}_1 \cdot \text{height}_1 = \text{base}_2 \cdot \text{height}_2$$
$$10 \cdot h_1 = 7 \cdot h_2$$
$$\frac{h_1}{h_2} = \frac{7}{10}$$

The ratio, $h_1 : h_2$ equals $7 : 10$. Note that although you can determine the ratio of the heights, determining h_1 or h_2 or the area of $PQRS$ is impossible.

2 AM is 15, so $\triangle ARM$ is in the $3 : 4 : 5$ family of triangles. In fact, it is a $3 - 4 - 5$ triangle blown up five times. $\overline{RM}$ is the long leg and is thus 4 times 5, or 20 units long.

$$\text{Area}_{GRAM} = \text{base} \cdot \text{height}$$
$$= AM \cdot RM$$
$$= 15 \cdot 20$$
$$= 300 \text{ units}^2$$

GM is 25, so

$$\text{Area}_{GRAM} = \text{base} \cdot \text{height}$$
$$300 = GM \cdot h$$
$$300 = 25 \cdot h$$
$$12 = h$$

3 To get h, you need to use the trapezoid area formula, and to use the formula, you need the sum of the lengths of the bases, $\overline{XY}$ and $\overline{WZ}$. (Note that all you need is the sum; you don't need to know the lengths of the individual bases.)

$$XY + WZ + 6 + 7 = 35$$
$$XY + WZ = 22$$

$$\text{Area}_{WXYZ} = \frac{b_1 + b_2}{2} \cdot h$$
$$55 = \frac{XY + WZ}{2} \cdot h$$
$$55 = \frac{22}{2} \cdot h$$
$$\frac{55}{11} = h$$
$$5 = h$$

(*4) The heights of the three sideways trapezoids and the triangle in quadrant I are all equal to 2 (note that the heights run along the x-axis). You need the bases of the trapezoids and the triangle to compute their areas, and their bases equal the y-coordinates at $x = 0$, 2, 4, and 6. When x is 0, y is 8, so that left-most base is 8. To find the other bases, plug the x-coordinates into the equation of the circle. Thus, you find that

$$x^2 + y^2 = 64 \qquad\qquad 4^2 + y^2 = 64 \qquad\qquad 6^2 + y^2 = 64$$
$$2^2 + y^2 = 64 \qquad\qquad\qquad y = 4\sqrt{3} \qquad\qquad\qquad y = 2\sqrt{7}$$
$$y^2 = 60$$
$$y = 2\sqrt{15}$$

So the first trapezoid (between $x = 0$ and $x = 2$) has bases of 8 and $2\sqrt{15}$, the second trapezoid has bases of $2\sqrt{15}$ and $4\sqrt{3}$, and the third has bases of $4\sqrt{3}$ and $2\sqrt{7}$. The single triangle's base is $2\sqrt{7}$. Compute their areas, add them up, and then multiply that result by 4 (for the four quadrants):

$$\text{Area}_{\text{1st Trap}} = \frac{b_1 + b_2}{2} \cdot h \qquad\qquad\qquad \text{Area}_{\text{3rd Trap}} = \frac{4\sqrt{3} + 2\sqrt{7}}{2} \cdot 2$$
$$= \frac{8 + 2\sqrt{15}}{2} \cdot 2 \qquad\qquad\qquad\qquad = 4\sqrt{3} + 2\sqrt{7}$$
$$= 8 + 2\sqrt{15}$$

$$\text{Area}_{\triangle} = \frac{1}{2} \cdot \text{base} \cdot \text{height}$$
$$\text{Area}_{\text{2nd Trap}} = \frac{2\sqrt{15} + 4\sqrt{3}}{2} \cdot 2 \qquad\qquad = \frac{1}{2} \cdot 2\sqrt{7} \cdot 2$$
$$= 2\sqrt{15} + 4\sqrt{3} \qquad\qquad\qquad\qquad = 2\sqrt{7}$$

$$\text{Total of four areas} = \left(8 + 2\sqrt{15}\right) + \left(2\sqrt{15} + 4\sqrt{3}\right) + \left(4\sqrt{3} + 2\sqrt{7}\right) + 2\sqrt{7}$$
$$= 8 + 4\sqrt{15} + 8\sqrt{3} + 4\sqrt{7}$$

$$\text{Estimate of circle's area} = 4\left(8 + 4\sqrt{15} + 8\sqrt{3} + 4\sqrt{7}\right)$$
$$= 32 + 16\sqrt{15} + 32\sqrt{3} + 16\sqrt{7}$$
$$\approx 191.7 \text{ units}^2$$

In case you're curious, the area of the circle is

$$\text{Area}_{\text{Circle}} = \pi r^2$$
$$= \pi \cdot 8^2$$
$$= 64\pi$$
$$\approx 201.1 \text{ units}^2$$

The estimate was a little less than 5 percent off.

(*5) A good plan of attack here — like with so many polygon problems — is to cut the figure up into right triangles and rectangles. Draw altitudes from K to $\overline{JM}$ and from L to $\overline{JM}$, and call their length h. On the left, you have a $30°-60°-90°$ triangle with h as the long leg, so the short leg (along $\overline{JM}$) is $\dfrac{h}{\sqrt{3}}$ (because in a $30°-60°-90°$ triangle, the ratio of short leg to long leg is $1:\sqrt{3}$ — see Chapter 4 for details). On the right, you have another $30°-60°-90°$ triangle, but this time h is the short leg. The long leg (along $\overline{JM}$) is thus $h\sqrt{3}$. KL is 10, so the distance between the two altitudes along $\overline{JM}$ is also 10. So, you have three pieces along $\overline{JM}$ that add up to 18. Now you can find h:

$$\frac{h}{\sqrt{3}} + 10 + h\sqrt{3} = 18$$
$$\sqrt{3}\left(\frac{h}{\sqrt{3}} + 10 + h\sqrt{3}\right) = 18 \cdot \sqrt{3}$$
$$h + 10\sqrt{3} + 3h = 18\sqrt{3}$$
$$4h = 8\sqrt{3}$$
$$h = 2\sqrt{3}$$

Now that you know h, simply plug it into the trapezoid area formula along with the given bases:

$$\text{Area}_{JKLM} = \frac{b_1 + b_2}{2} \cdot h$$
$$= \frac{10 + 18}{2} \cdot 2\sqrt{3}$$
$$= 28\sqrt{3} \text{ units}^2$$

(6) Draw diagonal $\overline{BD}$, which is the perpendicular bisector of $\overline{AC}$, and label the intersection of the diagonals X. Triangle ABC is equilateral, so AC is 10; thus, AX and XC are each 5. Triangle ABX is a $30°-60°-90°$ triangle with a short leg of 5, so its long leg, $\overline{BX}$, has a length of $5\sqrt{3}$. Triangle XCD is a $5-12-13$ right triangle with XD equal to 12 (see Chapter 4 for more on Pythagorean triples). Thus, $\overline{BD}$ has a length of $12 + 5\sqrt{3}$. Finally,

$$\text{Area}_{ABCD} = \frac{1}{2}d_1 d_2$$
$$= \frac{1}{2}AC \cdot BD$$
$$= \frac{1}{2} \cdot 10\left(12 + 5\sqrt{3}\right)$$
$$= 60 + 25\sqrt{3} \text{ units}^2$$

(7) Draw diagonal $\overline{RU}$, which is the perpendicular bisector of $\overline{BS}$, and label the intersection of the diagonals X. That step creates a right triangle with a leg of 8 and a hypotenuse of 15. Careful now — this is *not* an $8-15-17$ right triangle (remember, the hypotenuse is the longest side). Use the Pythagorean Theorem to get XU:

$$XU^2 + XS^2 = US^2$$
$$XU^2 + 8^2 = 15^2$$
$$XU^2 = 161$$
$$XU = \sqrt{161}$$

RU is twice *XU* because the diagonals in a rhombus bisect each other, so *RU* is $2\sqrt{161}$; using the kite formula,

$$\text{Area}_{RBUS} = \frac{1}{2}d_1d_2$$
$$= \frac{1}{2} \cdot 16 \cdot 2\sqrt{161}$$
$$= 16\sqrt{161}$$

8) ∠*S* and ∠*T* are supplementary (property of a parallelogram — see Chapter 6), so ∠*S* is 60°, and Δ*RSZ* is thus a 30°–60°–90° triangle with a long leg of 4. The short leg, $\overline{SZ}$, therefore measures $\frac{4}{\sqrt{3}}$, and the hypotenuse, $\overline{RS}$, is twice that, or $\frac{8}{\sqrt{3}}$ units long. All sides of a rhombus are equal, so *ST* is also $\frac{8}{\sqrt{3}}$. This time, you use the parallelogram formula to get the desired area:

$$\text{Area}_{QRST} = \text{base} \cdot \text{height}$$
$$= ST \cdot RZ$$
$$= \frac{8}{\sqrt{3}} \cdot 4$$
$$= \frac{32}{\sqrt{3}}$$
$$\approx 18.5 \text{ units}^2$$

9) If you slide the span up and to the right till it hits the center, you can see that the span is twice the apothem, so the apothem is 16. And as you can see in the example problem, a regular hexagon consists of six equilateral triangles, and its apothem is the altitude of one of these equilateral triangles. See the following figure:

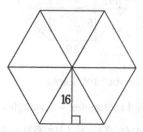

This apothem is the long leg of a 30°–60°–90° triangle, so the short leg is $\frac{16}{\sqrt{3}}$, and the hypotenuse is twice that, or $\frac{32}{\sqrt{3}}$ (the ratio of sides in a 30°–60°–90° triangle is $1 : \sqrt{3} : 2$, see Chapter 4 for details). And that hypotenuse is a side of the equilateral triangle. Each side of the hexagon is, therefore, $\frac{32}{\sqrt{3}}$, and the perimeter is six times that, or $\frac{192}{\sqrt{3}}$. You're finally ready to use the area formula:

$$\text{Area}_{\text{Reg. Poly.}} = \frac{1}{2}pa$$
$$= \frac{1}{2} \cdot \frac{192}{\sqrt{3}} \cdot 16$$
$$= \frac{1536}{\sqrt{3}}$$
$$= 512\sqrt{3} \text{ units}^2$$

10 Find the right lines to draw? Here they are:

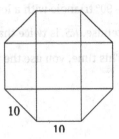

The four triangles are $45° - 45° - 90°$ triangles with a hypotenuse of 10. These triangles have sides with lengths in the ratio $1:1:\sqrt{2}$ (see Chapter 4 for more information). The legs of the $45° - 45° - 90°$ triangles are thus $\frac{10}{\sqrt{2}}$, or $5\sqrt{2}$. So now you have

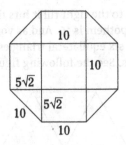

Now add up all the pieces for your total area:

$$\text{Area}_{\text{Octagon}} = 1 \text{ square} + 4 \text{ rectangles} + 4 \text{ triangles}$$
$$= 10^2 + 4(10)\left(5\sqrt{2}\right) + 4\left(\frac{1}{2}\right)\left(5\sqrt{2}\right)\left(5\sqrt{2}\right)$$
$$= 100 + 4\left(50\sqrt{2}\right) + 4(25)$$
$$= 200 + 200\sqrt{2} \text{ units}^2$$

11 *TAYLOR* is a hexagon, so the sum of its interior angles is $(n-2)180 = (6-2)180 = 720°$. Subtract the four known angles from this value: $720 - (170 + 95 + 161 + 100) = 194°$. Then, because the two remaining angles, $\angle ATR$ and $\angle O$, are congruent, each must be half of $194°$, or $97°$. Finally, because $\angle 1$ is the supplement of $\angle ATR$, $\angle 1 = 180 - 97 = 83°$.

(12) Here you go:

a. $180(n-2) = 1080$
$$n - 2 = 6$$
$$n = 8 \text{ sides}$$

b. $180(n-2) = 7920$
$$n - 2 = 44$$
$$n = 46 \text{ sides}$$

c. $180(n-2) = 180x^2 + 180$
$$n - 2 = x^2 + 1$$
$$n = x^2 + 3 \text{ sides}$$

d. $180(n-2) = 825$
$$n - 2 = 4.58$$
$$n = 6.58 \text{ sides}$$

There's no such thing as a polygon with 6.58 sides.

(13) Here are the angle measures:

a. The total is 360°. One exterior angle is $360 \div 5 = 72°$.

b. The total is 360°. One exterior angle is $360 \div 18 = 20°$.

c. The total is 360°. You can't compute the measure of a single exterior angle because you don't know whether the icosagon is equiangular.

d. The total is 360°. The fact that the 80-gon is equilateral does not tell you whether it's equiangular, so you can't figure the measure of a single exterior angle.

(14) Number of diagonals $= \dfrac{n(n-3)}{2}$
$$= \dfrac{30(30-3)}{2}$$
$$= 405$$

(15) Number of diagonals $= 3 \cdot (\text{number of sides})$
$$\dfrac{n(n-3)}{2} = 3n$$
$$n^2 - 3n = 6n$$
$$n^2 - 9n = 0$$
$$n(n-9) = 0$$
$$n = 0 \text{ or } 9$$

(16) There's no such thing as a polygon with zero sides, so the answer is 9.

$$\dfrac{n(n-3)}{2} = 54$$
$$n^2 - 3n = 108$$
$$n^2 - 3n - 108 = 0$$
$$(n-12)(n+9) = 0$$
$$n = 12 \text{ or } -9$$

So, you have an equiangular 12-gon; therefore,

$$\text{One interior angle} = 180° - \dfrac{360°}{n}$$
$$= 180° - \dfrac{360°}{12}$$
$$= 150°$$

Chapter **8**

Similarity: Size Doesn't Matter (Including Proofs)

When two triangles, two rectangles, two pentagons, or two of any type of polygon have the same shape (regardless of whether they have the same size), you say that they're *similar* —like if you take a figure and blow it up or shrink it down in a photocopy machine, the new image will be the same shape as (and thus *similar* to) the original. Congruent figures are automatically similar, but when you do problems involving two similar figures, you're usually dealing with two things of different sizes that have the same shape. The squiggle symbol, ~, means *is similar to*.

In this chapter, you do problems involving similar triangles and similar polygons of more than three sides. Similar polygons have proportional sides, so you also do many problems involving proportions. Finally, you practice using theorems — some of which have nothing to do with similarity — that, like similarity theorems, also involve proportions.

Defining Similarity

When you're talking about similarity, you have to talk about the two defining characteristics of similar figures.

REMEMBER

Similar polygons: In similar polygons, both of the following are true.

>> **Corresponding angles are congruent.** If objects of different sizes have the same shape, their angles have to be equal. This idea is kinda obvious if you think about it. Imagine you see something like a yield sign on the side of the road. It's a downward-pointing equilateral triangle with three 60° angles. As you get closer, it looks bigger, of course, but regardless of how big or small it looks, the three angles are always 60° angles. If the angles were to change to something other than 60°, the sign would no longer look like a yield sign. It would've morphed into a different shape.

>> **The ratios of the lengths of corresponding sides are equal.** Say the front door of a house is 7 feet tall and 3 feet wide and that the blueprint for the design of the house contains a door measuring 2.1 inches tall by 0.9 inches wide. Because these two doors are similar rectangles, the ratio of their heights equals the ratio of their widths, and you get the following proportion:

$$\frac{\text{height}_{\text{real door}}}{\text{height}_{\text{blueprint door}}} = \frac{\text{width}_{\text{real door}}}{\text{width}_{\text{blueprint door}}}$$

$$\frac{7\,\text{feet}}{2.1\,\text{inches}} = \frac{3\,\text{feet}}{0.9\,\text{inch}}$$

Both sides of the second equation equal 40, which tells you that the real door is 40 times as tall and 40 times as wide as the door shown in the blueprint (you have to convert all units to inches or feet before calculating this). Such ratios or quotients represent the blow-up or shrink factor, depending on which way you look at it.

Perimeters of similar polygons: The ratio of the perimeters of two similar polygons equals the ratio of any pair of corresponding sides.

THEOREMS & POSTULATES

EXAMPLE

Q. Given: Pentagon *ABCDE* ~ pentagon *VWXYZ*

Perimeter of *ABCDE* is 18

Find:

a. *VW*

b. *XY*

c. Perimeter of *VWXYZ*

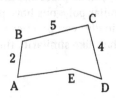

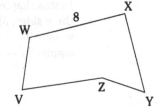

A. In the diagram, the two pentagons have the same *orientation*; in other words, *A* matches up with *V*, *B* matches up with *W*, and so on. If you were to expand *ABCDE* a bit and slide it over to the right, it would fit perfectly on top of *VWXYZ*. You wouldn't have to rotate it or flip it upside down to make it fit. But this isn't always the case, so to make sure you're pairing up the correct vertices and sides, pay attention to the way the similarity is written. When someone says *ABCDE ~ VWXYZ*, it means that *A* pairs up with *V*, *B* pairs up with *W*, and so on, and that $\overline{CD}$ (the third and fourth letters) pairs up with $\overline{XY}$ (also the third and fourth letters). Got it? Fantastic!

a. $\dfrac{\text{left side}_{VWXYZ}}{\text{left side}_{ABCDE}} = \dfrac{\text{top}_{VWXYZ}}{\text{top}_{ABCDE}}$

$$\frac{VW}{2} = \frac{8}{5}$$

$$VW = \frac{16}{5} = 3.2$$

b. $\dfrac{\text{right side}_{VWXYZ}}{\text{right side}_{ABCDE}} = \dfrac{\text{top}_{VWXYZ}}{\text{top}_{ABCDE}}$

$$\frac{XY}{2} = \frac{8}{5}$$

$$XY = \frac{32}{5} = 6.4$$

(Or you could just notice that the right side of *ABCDE* is twice as long as the left side, so you can simply multiply *VW* by 2 to get *XY*.)

c. $\dfrac{\text{perimeter}_{VWXYZ}}{\text{perimeter}_{ABCDE}} = \dfrac{\text{side}_{VWXYZ}}{\text{side}_{ABCDE}}$

$$\frac{\text{perimeter}_{VWXYZ}}{18} = \frac{8}{5}$$

$$\text{perimeter}_{VWXYZ} = \frac{144}{5} = 28.8$$

 Given: *ABCD ~ EFGH*

Find: **a.** All missing angles

        b. All missing sides

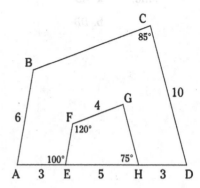

Given: △PQR ~ △ZXY

Find: **a. All missing angles**
         b. All missing sides

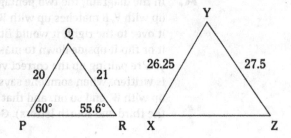

3

Given: ABCDE ~ LMNOP

         Perimeter of ABCDE is 30

Find: Perimeter of LMNOP

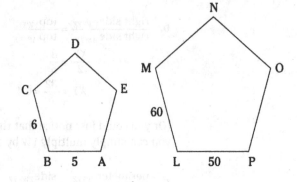

4

Given: △ABD ~ △ACD

Find: **a.** AD
         b. DB

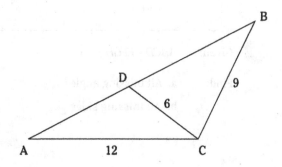

Proving Triangles Similar

You have five ways to prove triangles congruent: SSS, SAS, ASA, AAS, and HLR (see Chapter 5). Now you get three ways to prove triangles similar: SSS~, SAS~, and AA. The most frequently used and by far the easiest to use is AA.

Proving triangles similar:

THEOREMS & POSTULATES

>> **AA.** If two angles of one triangle are congruent to two angles of another triangle, then the triangles are similar.

>> **SSS~.** If the ratios of the three pairs of corresponding sides of two triangles are equal, then the triangles are similar.

>> **SAS~.** If the ratios of two pairs of corresponding sides of two triangles are equal and the included angles are congruent, then the triangles are similar.

EXAMPLE

Q. Given: $\angle 1 \cong \angle 3$

$\angle 2 \cong \angle 4$

Prove: $\triangle BLO \sim \triangle WUP$

Find: WU

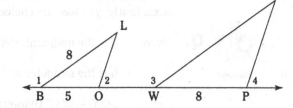

A.

Statements	Reasons
1) $\angle 1 \cong \angle 3$	1) Given.
2) $\angle LBO \cong \angle UWP$	2) Supplements of congruent angles are congruent.
3) $\angle 2 \cong \angle 4$	3) Given.
4) $\angle LOB \cong \angle UPW$	4) Supplements of congruent angles are congruent.
5) $\triangle BLO \sim \triangle WUP$	5) AA (if two angles of one triangle are congruent to two angles of another triangle, then the triangles are similar).

Now find WU. Piece o' cake:

$$\frac{\text{top}_{\triangle WUP}}{\text{top}_{\triangle BLO}} = \frac{\text{base}_{\triangle WUP}}{\text{base}_{\triangle BLO}}$$

$$\frac{WU}{8} = \frac{8}{5}$$

$$WU = \frac{64}{5} = 12.8$$

Q. Given: Diagram as shown

EXAMPLE

Prove: △MAR ~ △BLE (paragraph proof)

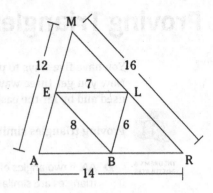

A. No reason to bother with a two-column proof here. All you have to do is to show that all three ratios of corresponding sides are equal, like this: $\frac{12}{6} = \frac{14}{7} = \frac{16}{8}$. Check.

Thus, by SSS~, the triangles are similar. But you still have to show that the right vertices pair up. One way to do this is to pick a vertex, like A, and note that it's across from the longest side of △MAR (16). So, it corresponds to L, which is across from the longest side of △BLE (8). Then R and E correspond because both are across from the shortest sides. Lastly, you have no choice, of course, but to pair M with B. Thus, △MAR ~ △BLE.

Q. Given: U is the midpoint of $\overline{RA}$

EXAMPLE

G is the midpoint of $\overline{RT}$

Prove: △RUG ~ △RAT (paragraph proof)

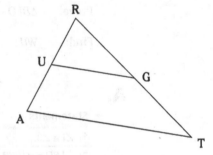

A. Let's skip the two-column mumbo jumbo again. (You can do this proof in two-column format, but that involves all sorts of rigamarole like 1) U is a midpoint; then 2) $\overline{RU} \cong \overline{UA}$; then 3) $RU = UA$; then 4) $RU + UA = RA$; then 5) $RU + RU = RA$; then 6) $2 \cdot RU = RA$; then 7) $\frac{RA}{RU} = 2$; and so on, and so on.) So, just use common sense instead. Because U is the midpoint of $\overline{RA}$, you know that $\frac{RU}{RA} = \frac{1}{2}$. G works the same way, so $\frac{RG}{RT} = \frac{1}{2}$. Thus, $\frac{RU}{RA} = \frac{RG}{RT}$. Then, because $\angle R = \angle R$, △RUG ~ △RAT by SAS~.

180 PART 3 Polygons, Proof and Non-Proof Problems

5 Given: Diagram as shown

Prove: $\triangle PQR \sim \triangle STU$ (paragraph proof)

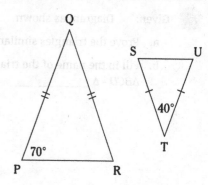

 6 Given: $\triangle XRT$ is isosceles with base $\overline{XT}$ and altitude $\overline{RN}$

   $\angle 1 \cong \angle 2$

Prove: $\triangle WGN \sim \triangle TRN$

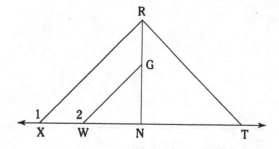

Statements	Reasons

Given: Diagram as shown

    a. Prove the triangles similar (paragraph proof)

    b. Fill in the name of the triangle:
       $\triangle BCD \sim \triangle$ _____ .

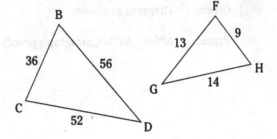

8 Given: $\overline{AT} \parallel \overline{OY}$

Prove: $\triangle BOY$ is isosceles (paragraph proof)

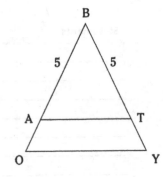

Given: $\angle ELP \cong \angle IPL$

$\angle EPL \cong \angle ILP$

V is the midpoint of $\overline{LI}$

S is the midpoint of $\overline{PI}$

Prove: $\triangle VIS \sim \triangle PEL$

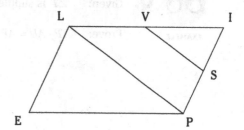

Statements	Reasons

Corresponding Sides and CSSTP — Cats Stalk Silently Then Pounce

Actually, CSSTP stands for *Corresponding Sides of Similar Triangles are Proportional.* You can tell this statement is true from the definition of similar polygons. And if you've done the preceding problems, you've used this concept already when you had to calculate the lengths of the sides of similar triangles and other polygons. What's new here is using CSSTP in formal, two-column proofs.

You use CSSTP on the line immediately after showing triangles similar, just like you use CPCTC on the line after you show triangles congruent (for more on congruent parts of congruent triangles, see Chapter 5).

TIP

Q. Given: ∠T is supplementary to ∠UAC

Prove: $TP \cdot AU = AP \cdot TC$

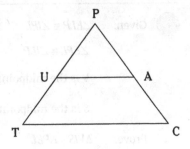

A. *Tip:* When you're asked to prove that a product equals another product (like $TP \cdot AU = AP \cdot TC$ in this example proof), the proof likely involves similar triangles (or perhaps — though less likely — one of the three theorems in the upcoming section). So, look for similar-looking triangles that contain the four segments in the *prove* statement.

Statements	Reasons
1) ∠T is supplementary to ∠UAC	1) Given.
2) ∠PAU is supplementary to ∠UAC	2) Two angles that form a straight angle are supplementary.
3) ∠T ≅ ∠PAU	3) Supplements of the same angle are congruent.
4) ∠P ≅ ∠P	4) Reflexive Property.
5) △TPC ~ △APU	5) AA (Statements 3 and 4).
6) $\frac{TP}{AP} = \frac{TC}{AU}$	6) CSSTP.
7) $TP \cdot AU = AP \cdot TC$	7) Means-Extremes Products Theorem (a fancy name for cross-multiplication).

10 Given: ∠JKL ≅ ∠NML

Perimeter of △JKL is 27

Prove: $\overline{MN} \cong \overline{JL}$ (paragraph proof)

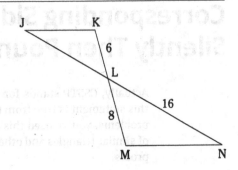

11 Given: $\angle 1 \cong \angle CBD$

Prove: $(AD)^2 = (AB)(AC)$ with a paragraph proof

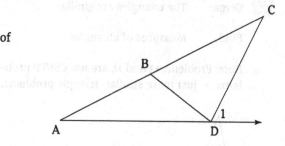

*12 Given: E and O trisect $\overline{AS}$

O bisects $\overline{RM}$

$\angle S \cong \angle MTI$

Prove: $RO \cdot IT = MI \cdot OE$

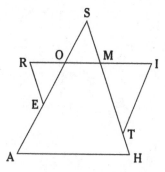

Statements	Reasons

13 Given: The triangles are similar

Find: Measures of all angles

Note: Problems 13 and 14 are not CSSTP problems — just more similar-triangle problems.

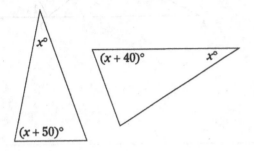

14 Indicate whether statements **a** through **f** are *always* true, *sometimes* true, or *never* true.

a. If △*ABC* ~ △*CBA*, then $\overline{AB} \cong \overline{CB}$.

b. If ∠*ABC* ≅ ∠*DEF*, then △*ABC* ~ △*DEF*.

c. If △*ABC* ≅ △*DEF*, then △*ABC* ~ △*DEF*.

d. If △*ABC* ~ △*DEF*, then △*ABC* ≅ △*DEF*.

e. If △*ABC* is a right triangle and △*DEF* is an acute triangle, then △*ABC* ~ △*DEF*.

f. If △*ABC* and △*DEF* are both isosceles and ∠*B* ≅ ∠*E*, then △*ABC* ~ △*DEF*.

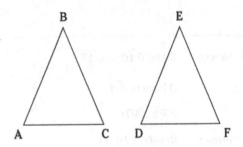

Similar Rights: The Altitude-on-Hypotenuse Theorem

If you use the hypotenuse of a right triangle as its base and draw an altitude to it — creating two more, smaller right triangles — all three triangles are similar. Here's the handy-dandy theorem.

Altitude-on-Hypotenuse Theorem: If an altitude is drawn to the hypotenuse of a right triangle as shown in Figure 8-1, then

>> The two triangles formed are similar to the given triangle and to each other:

$$\triangle ACB \sim \triangle ADC \sim \triangle CDB$$

>> $h^2 = xy$

>> $a^2 = yc$ and $b^2 = xc$ (note that this is really just one formula or relationship, not two. It works exactly the same on both sides of the big triangle):

$$(\text{leg of big } \triangle)^2 = (\text{part of hypotenuse next to leg}) \cdot (\text{whole hypotenuse})$$

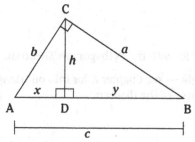

FIGURE 8-1:
Three similar right triangles in one: Triple the pleasure, triple the fun.

TIP

There's more than one way to peel an orange. When doing a problem involving an altitude-on-hypotenuse diagram (like Figure 8-1), don't assume that the problem must be solved with the second or third part of the Altitude-on-Hypotenuse Theorem. Sometimes, the easiest way to solve the problem is with the Pythagorean Theorem. And at other times, you can use ordinary similar-triangle proportions to solve the problem.

EXAMPLE

Q. Use the figure to answer the following questions.

a. If $PQ = 12$ and $QR = 3$, find QS, PS, and RS

b. If $PR = 13$ and $RS = 5$, find PS, PQ, QR, and QS

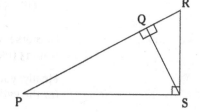

A. Here's how this problem plays out:

a. From the second part of the theorem, $h^2 = xy$, so

$$(QS)^2 = (PQ)(QR)$$
$$= 12 \cdot 3$$
$$= 36$$
$$QS = 6$$

From the third part of the theorem, $a^2 = yc$ and $b^2 = xc$, so

$$(PS)^2 = (PQ)(PR)$$
$$= 12 \cdot 15$$
$$= 180$$
$$PS = \sqrt{180}$$
$$= 6\sqrt{5}$$

and

$$(RS)^2 = (QR)(PR)$$
$$= 3 \cdot 15$$
$$= 45$$
$$RS = \sqrt{45}$$
$$= 3\sqrt{5}$$

Of course, you could also get PS and RS with the Pythagorean Theorem.

b. PS is 12 (you have a $5-12-13$ triangle — see Chapter 4 for info on triangle families). You can get PQ and QR using part three of the theorem:

$$(PS)^2 = (PQ)(PR)$$
$$12^2 = PQ \cdot 13$$
$$PQ = \frac{144}{13}$$

and

$$(RS)^2 = (QR)(PR)$$
$$5^2 = QR \cdot 13$$
$$QR = \frac{25}{13}$$

Of course, you can just calculate one of these two lengths and then subtract it from 13 (PR) to get the other.

Finally, you get QS with the second part of the Altitude-on-Hypotenuse Theorem (or the Pythagorean Theorem):

$$(QS)^2 = (PQ)(QR)$$
$$= \left(\frac{144}{13}\right)\left(\frac{25}{13}\right)$$
$$= \frac{12^2 \cdot 5^2}{13^2}$$
$$QS = \frac{12 \cdot 5}{13} = \frac{60}{13}$$

15 Use the figure to calculate these lengths:

    a. If $JA = 4$ and $AY = 9$, find JZ and AZ

    b. If $JA = 3$ and $JZ = 5$, find AY

    c. If $JA = 2$ and $JY = 8$, find YZ

    d. If $AZ = 8$ and $AY = 10$, find JY

    e. If $JZ = 8$ and $JY = 12$, find AY

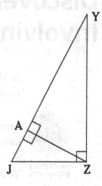

16 If $RQ = 5$ and $RS = 10$, find RT

    Hint: You can solve this by using the last two parts of the theorem, but there's an easier way.

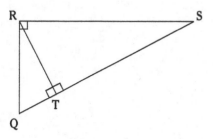

*17 Find FL

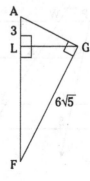

Discovering Three More Theorems Involving Proportions

In this last section, you practice using three more theorems that involve a proportion. The first two are related to similar triangles.

Side-Splitter Theorem: If a line is parallel to a side of a triangle and it intersects the other two sides, it divides those sides proportionally.

You can use the Side-Splitter Theorem *only* for the four segments on the split sides of the triangle. Do *not* use it for the parallel sides. For the parallel sides, use similar triangle proportions. (Whenever a triangle is divided by a line parallel to one of its sides, the small triangle created is similar to the original, large triangle. This idea follows from the *if two parallel lines are cut by a transversal, then corresponding angles are congruent* theorem [see Chapter 6] and AA.)

The theorem that shall not be named: If three or more parallel lines are intersected by two or more transversals, the parallel lines divide the transversals proportionally. Consider Figure 8-2. Given that the horizontal lines are parallel, the following proportions (among others), follow from the theorem:

$$\frac{AB}{CD} = \frac{PQ}{RS}, \quad \frac{AC}{CD} = \frac{WY}{YZ}, \quad \frac{PQ}{QS} = \frac{WX}{XZ}, \quad \frac{RS}{QR} = \frac{YZ}{XY}$$

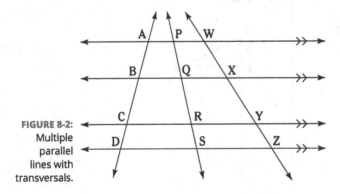

FIGURE 8-2: Multiple parallel lines with transversals.

The third theorem has nothing to do with similar figures. It's in this section because it involves a proportion.

Angle-Bisector Theorem: If a ray bisects an angle of a triangle, it divides the opposite side into segments that are proportional to the adjacent sides.

When you bisect an angle in a triangle, you *never* get similar triangles (except when you bisect the vertex angle of an isosceles triangle, in which case the resulting triangles are congruent as well as similar). The fact that the Angle-Bisector Theorem is usually in the similar triangle chapter in geometry books despite its having nothing to do with similar triangles may be one reason students often fail to remember the theorem.

Don't forget the Angle-Bisector Theorem. Whenever you see a triangle with one of its angles bisected, you either have an isosceles triangle cut into two congruent triangles or a problem in which you probably have to use the Angle-Bisector Theorem.

Q. Given that $\overline{YA} \parallel \overline{DN}$, find LA

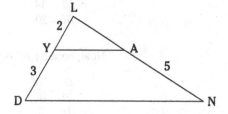

A. By the Side-Splitter Theorem,

$$\frac{LY}{YD} = \frac{LA}{AN}$$
$$\frac{2}{3} = \frac{LA}{5}$$
$$3 \cdot LA = 10$$
$$LA = \frac{10}{3} = 3\frac{1}{3}$$

Q. Given that $AC = 25$ and $FH = 36$, find BC, AB, IJ, GH, FG, and EF

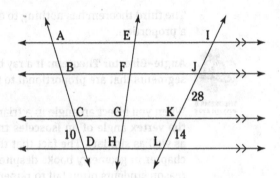

A. First, set up a proportion to find BC:

$$\frac{BC}{CD} = \frac{JK}{KL}$$

$$\frac{BC}{10} = \frac{28}{14}$$

$$BC = 20$$

Now, because AC is 25 (given), AB must be 5.

$$\frac{IJ}{JK} = \frac{AB}{BC}$$

$$\frac{IJ}{28} = \frac{5}{20}$$

$$IJ = 7$$

To get FG and GH, note that because the ratio $KL : JK$ is $14 : 28$ or $1 : 2$, $GH : FG$ must also equal $1 : 2$. So let $GH = x$ and $FG = 2x$. Then, because $FH = 36$ (given),

$$x + 2x = 36$$

$$x = 12$$

Therefore, GH is 12 and FG is 24. Finally,

$$\frac{EF}{FG} = \frac{AB}{BC}$$

$$\frac{EF}{24} = \frac{5}{20}$$

$$EF = 6$$

Q. Given: $\overrightarrow{AH}$ bisects $\angle CAS$

EXAMPLE

Find: *CH* and *HS*

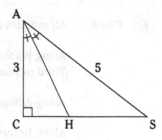

A. By the Angle–Bisector Theorem, $\dfrac{CH}{HS} = \dfrac{AC}{AS}$.

If you set *CH* equal to *x*, *HS* is 4 − *x*. (You saw that *CS* is 4, right?) Now substitute:

$$\frac{x}{4-x} = \frac{3}{5}$$
$$5x = 12 - 3x$$
$$8x = 12$$
$$x = \frac{12}{8} = 1.5$$

Thus, *CH* is 1.5 and *HS* is 2.5.

Warning: Don't make the mistake of thinking that when an angle in a triangle is bisected, the opposite side will also be cut exactly in half. You can see in this example that side $\overline{CS}$ is *not* bisected. The opposite side often comes very close to being bisected and it often *looks* bisected, but as a matter of fact, the opposite side is divided in half only when you bisect the vertex angle of an isosceles triangle.

---

18 Given: $\overline{OI} \parallel \overline{GN}$

a. Prove: $\triangle RIO \sim \triangle RNG$ (paragraph proof)

b. Find: *IN* and *GN*

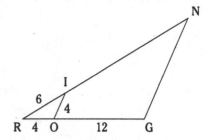

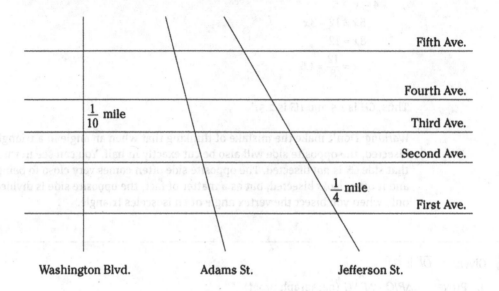

*19 Given: All avenues are parallel to one another.

Along Washington Blvd., it's $\frac{1}{2}$ of a mile from First Avenue to Fifth and $\frac{1}{10}$ of a mile from Third to Fourth.

Along Adams, it's $\frac{3}{8}$ of a mile from First to Third and from Third to Fifth.

Along Jefferson, it's $\frac{4}{5}$ of a mile from First to Fifth and $\frac{1}{4}$ of a mile from First to Second.

Find: All unknown distances between the avenues along Washington, Adams, and Jefferson, and fill in the distances in the following figure. (I've started it for you on Washington and Jefferson.) You need only the distances from one avenue to the next. In other words, you don't have to calculate the distance from Second to Fourth unless you need it to find one of the smaller distances.

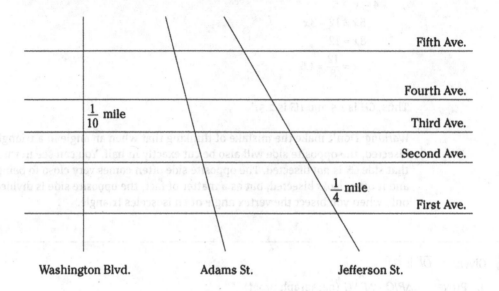

Solutions

1. Here are your answers:

   a. Corresponding angles of similar polygons are congruent, so finding the angles should be a cinch:

      - $\angle FEH$ is 80° and thus, so is $\angle A$

      - $\angle F = 120° = \angle B$

      - $\angle C = 85° = \angle G$

      - $\angle GHE = 75° = \angle D$

      - $\angle GHD = 105°$

   b. The ratio of the bases is $\frac{5}{11}$, so all the other ratios must also equal $\frac{5}{11}$:

   $$\frac{FE}{6} = \frac{5}{11} \qquad \text{Then, } \frac{4}{BC} = \frac{5}{11} \qquad \text{Finally, } \frac{GH}{10} = \frac{5}{11}$$
   $$FE = \frac{30}{11} \qquad\qquad BC = \frac{44}{5} \qquad\qquad GH = \frac{50}{11}$$

2. Here are the missing angles and lengths:

   a. The angles in $\triangle PQR$ must add up to 180°, so $\angle Q$ is 64.4°. Now just pair up corresponding vertices: P with Z, Q with X, and R with Y. Thus, $\angle Z$ is 60°, $\angle X$ is 64.4°, and $\angle Y$ is 55.6°.

   b. $\overline{QR}$ (second and third letters) corresponds to $\overline{XY}$ (second and third letters) and $\overline{PQ}$ corresponds to $\overline{ZX}$, so

   $$\frac{QR}{XY} = \frac{PQ}{ZX}$$
   $$\frac{21}{26.25} = \frac{20}{ZX}$$
   $$ZX = \frac{20 \cdot 26.25}{21} = 25$$

   $\overline{PR}$ corresponds to $\overline{ZY}$, so

   $$\frac{PR}{27.5} = \frac{21}{26.25}$$
   $$PR = \frac{27.5 \cdot 21}{26.25} = 22$$

3. Did you notice the trick? The base $\overline{AB}$ does *not* correspond to the base $\overline{LP}$. Base $\overline{AB}$ corresponds to $\overline{LM}$. So, the expansion factor is $\frac{60}{5} = 12$, not $\frac{50}{5} = 10$. Thus,

   $$\frac{\text{perimeter}_{LMNOP}}{\text{perimeter}_{ABCDE}} = \frac{LM}{AB}$$

   $$\frac{\text{perimeter}_{LMNOP}}{30} = \frac{60}{5}$$

   $$\text{perimeter}_{LMNOP} = 30 \cdot 12 = 360$$

(4) And this is how problem 4 plays out:

a. From the way the similarity is written, you can see that $\overline{AC}$ (first and third letters in $\triangle ABC$) pairs up with $\overline{AD}$ (first and third letters in $\triangle ACD$) and that $\overline{BC}$ pairs up with $\overline{CD}$. This gives you the desired proportion:

$$\frac{AD}{AC} = \frac{CD}{BC}$$

$$\frac{AD}{12} = \frac{6}{9}$$

$$AD = \frac{72}{9} = 8$$

b. Another way to see how things pair up is to redraw the triangles so they're side by side and in the same orientation. Like this:

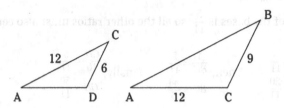

You can see that $\triangle ACD$ had to be flipped over to put it in the same orientation as $\triangle ABC$.

To get DB, you first need AB:

$$\frac{\text{top}_{\triangle ABC}}{\text{top}_{\triangle ACD}} = \frac{\text{right side}_{\triangle ABC}}{\text{right side}_{\triangle ACD}}$$

$$\frac{AB}{AC} = \frac{CB}{DC}$$

$$\frac{AB}{12} = \frac{9}{6}$$

$$AB = \frac{108}{6} = 18$$

Finally, you can see in the original figure that $DB = AB - AD$, so $DB = 18 - 8 = 10$.

(5) You know that $\angle R \cong \angle P$ by *if sides, then angles* (Chapter 5), so $\angle R$ is 70°. The angles in a triangle add up to 180°, so $\angle Q$ is 40°. You know $\angle S$ and $\angle T$ are also equal by *if sides, then angles*, and together they must sum to 140°, so each is 70°. Because both triangles contain a 40° angle and a 70° angle, $\triangle PQR \sim \triangle STU$ by AA.

After finding that $\angle Q$ is 40°, you could also finish by setting PQ and RQ equal to x and setting ST and UT equal to y. That gives you $\frac{PQ}{ST} = \frac{RQ}{UT}$ (because $\frac{x}{y} = \frac{x}{y}$).

And because $\angle Q \cong \angle T$, $\triangle PQR \sim \triangle STU$ by SAS~.

6) Statements	Reasons
1) $\triangle XRT$ is isosceles with base $\overline{XT}$	1) Given.
2) $\overline{XR} \cong \overline{TR}$	2) Definition of an isosceles triangle.
3) $\angle NXR \cong \angle NTR$	3) If sides, then angles.
4) $\angle 1 \cong \angle 2$	4) Given.
5) $\angle NXR \cong \angle NWG$	5) Supplements of congruent angles are congruent.
6) $\angle NWG \cong \angle NTR$	6) Transitive Property (Statements 3 and 5).
7) $\overline{RN}$ is an altitude	7) Given.
8) $\overline{RN} \perp \overline{XT}$	8) Definition of altitude.
9) $\angle GNW$ is a right angle $\angle RNT$ is a right angle	9) Definition of perpendicular.
10) $\angle GNW \cong \angle RNT$	10) Right angles are congruent.
11) $\triangle WGN \sim \triangle TRN$	11) AA (Statements 6 and 10).

7) Here's how this problem unfolds:

a. All you have to check is whether

$$\frac{\text{short side}_{\Delta 1}}{\text{short side}_{\Delta 2}} \overset{?}{=} \frac{\text{medium side}_{\Delta 1}}{\text{medium side}_{\Delta 2}} \overset{?}{=} \frac{\text{long side}_{\Delta 1}}{\text{long side}_{\Delta 2}}$$

$$\frac{36}{9} \overset{?}{=} \frac{52}{13} \overset{?}{=} \frac{56}{14}$$

$4 = 4 = 4$. Check.

b. B and H correspond because each is across from a medium-length side. C and F correspond because each is across from a long side. D and G are stuck with each other. Therefore, $\triangle BCD \sim \triangle HFG$.

8) Because $\overline{AT}$ is parallel to $\overline{OY}$, $\angle BAT \cong \angle BOY$ by *if parallel lines are cut by a transversal, then corresponding angles are congruent* (Chapter 6). Then, because both triangles contain $\angle B$, $\triangle BAT \sim \triangle BOY$ by AA. Similar triangles have proportional sides, so

$$\frac{BA}{BO} = \frac{BT}{BY}$$

$$\frac{5}{BO} = \frac{5}{BY}$$

$$5 \cdot BY = 5 \cdot BO$$

$$BY = BO$$

Thus, $\overline{BY} \cong \overline{BO}$

And, therefore, $\triangle BOY$ is isosceles.

After showing the triangles similar, you can also reason that any triangle similar to an isosceles triangle must also be isosceles, because if you take an isosceles triangle and shrink it or expand it by some factor, the two equal sides remain equal sides.

(9)

Statements	Reasons
1) V is the midpoint of $\overline{LI}$ S is the midpoint of $\overline{PI}$	1) Given.
2) $VI = \frac{1}{2} LI$ $SI = \frac{1}{2} PI$	2) A midpoint divides a segment into two segments that are each half as long as the original segment. (This reason is true, of course, but I created this theorem to avoid having to go through the rigamarole I referred to in the last example in this section.)
3) $\frac{VI}{LI} = \frac{1}{2}$; $\frac{SI}{PI} = \frac{1}{2}$	3) Algebra.
4) $\frac{VI}{LI} = \frac{SI}{PI}$	4) Substitution.
5) $\angle I \cong \angle I$	5) Reflexive Property.
6) $\triangle VIS \sim \triangle LIP$	6) SAS~ (Statements 4 and 5).
7) $\angle ELP \cong \angle IPL$	7) Given.
8) $\angle EPL \cong \angle ILP$	8) Given.
9) $\triangle PEL \sim \triangle LIP$	9) AA (Statements 7 and 8).
10) $\triangle VIS \sim \triangle PEL$	10) Transitive Property for similar triangles.

(10) The vertical angles are congruent and $\angle JKL \cong \angle NML$, so $\triangle JKL \sim \triangle NML$ by AA. Their sides are proportional; thus,

$$\frac{JL}{NL} = \frac{KL}{ML}$$

$$\frac{JL}{16} = \frac{6}{8}$$

$$JL = \frac{96}{8} = 12$$

Next, JK has to be 9 to make the perimeter of $\triangle JKL$ add up to 27. Finally,

$$\frac{MN}{JK} = \frac{ML}{KL}$$

$$\frac{MN}{9} = \frac{8}{6}$$

$$MN = \frac{72}{6} = 12$$

MN and JL are both 12, so of course, $\overline{MN} \cong \overline{JL}$.

11) You have to prove that a product equals another product — $(AD)^2$ is a product — so the tip in the example problem about looking for similar triangles applies. Thus, you look for triangles that contain AD, AB, and AC — namely $\triangle ABD$ and $\triangle ACD$ — and try to prove that they're similar. You know that $\angle 1 \cong \angle CBD$, so their supplements, $\angle ADC$ and $\angle ABD$, are congruent. Both triangles contain $\angle A$, so $\triangle ABD \sim \triangle ADC$ by AA. (Note the order of the vertices.) Now find a proportion that contains AB and AC and that uses AD twice, and you're done. Here it is:

$$\frac{\text{medium side}_{\triangle ABD}}{\text{medium side}_{\triangle ADC}} = \frac{\text{long side}_{\triangle ABD}}{\text{long side}_{\triangle ADC}}$$

$$\frac{AB}{AD} = \frac{AD}{AC}$$

$$(AD)^2 = (AB)(AC)$$

*12)

Statements	Reasons
1) E and O trisect $\overline{AS}$	1) Given.
2) $\overline{EO} \cong \overline{SO}$	2) Definition of trisect.
3) O bisects $\overline{RM}$	3) Given.
4) $\overline{RO} \cong \overline{MO}$	4) Definition of bisect.
5) $\angle ROE \cong \angle MOS$	5) Vertical angles are congruent.
6) $\triangle ROE \cong \triangle MOS$	6) SAS (2, 5, 4).
7) $\angle S \cong \angle MTI$	7) Given.
8) $\angle SMO \cong \angle TMI$	8) Vertical angles are congruent.
9) $\triangle MOS \sim \triangle MIT$	9) AA (Statements 7 and 8).
10) $\triangle ROE \sim \triangle MIT$	10) Substitution of $\triangle ROE$ for $\triangle MOS$ (Statements 6 and 9).
11) $\dfrac{RO}{MI} = \dfrac{OE}{IT}$	11) CSSTP.
12) $RO \cdot IT = MI \cdot OE$	12) Means-extremes (cross-multiplication).

13) The triangles are similar, so they must have three pairs of congruent angles according to the definition of similar polygons. Thus, each triangle must contain angles with measures x, $x + 40$, and $x + 50$. These must add up to $180°$, so

$$x + (x + 40) + (x + 50) = 180$$
$$3x + 90 = 180$$
$$3x = 90$$
$$x = 30$$

Thus, both triangles contain $30°$, $70°$, and $80°$ angles.

14) Here are the answers:

a. Always: You have only one triangle, so if $\triangle ABC \sim \triangle CBA$, $\triangle ABC$ must be congruent to $\triangle CBA$ as well. (Otherwise, $\triangle ABC$ and $\triangle CBA$ would be different sizes, which is impossible.) Vertex A (in $\triangle ABC$) corresponds to C (in $\triangle CBA$) and B corresponds to B, so $\overline{AB} \cong \overline{CB}$.

b. Sometimes: With only one pair of congruent angles, the triangles might be similar, but they don't have to be.

c. **Always:** Congruent triangles are automatically similar as well.

d. **Sometimes:** Similar triangles can be congruent, but they certainly don't have to be.

e. **Never:** $\triangle ABC$ contains a right angle. $\triangle DEF$ is acute, so it can't contain a right angle. Thus, the two triangles can't have three pairs of congruent angles, and therefore, they're not similar. (For more info on types of triangles, see Chapter 4.)

f. **Sometimes:** The answer would be always if you were told that $\overline{AC}$ and $\overline{DF}$ are the bases of these isosceles triangles, but the statement is only sometimes true because the triangles could look like this:

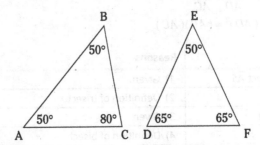

WARNING

Don't forget: **You can't rely on the appearance of the triangles in the figure** and conclude, for example, that $\overline{AB} \cong \overline{CB}$ and $\overline{DE} \cong \overline{FE}$.

15) Here are the lengths:

a. Using the third part of the Altitude-on-Hypotenuse Theorem,

$$(JZ)^2 = (JA)(JY)$$
$$= 4 \cdot 13$$
$$= 52$$
$$JZ = \sqrt{52} = 2\sqrt{13} \approx 7.2$$

Then, using the second part of the theorem,

$$(AZ)^2 = (JA)(AY)$$
$$= 4 \cdot 9$$
$$= 36$$
$$AZ = 6$$

b. Using the third part of the theorem,

$$(JZ)^2 = (JA)(JY)$$
$$5^2 = 3 \cdot JY$$
$$JY = \frac{25}{3} = 8\frac{1}{3}$$

Thus,

$$AY = JY - JA$$

$$= \frac{25}{3} - 3$$

$$= \frac{16}{3} = 5\frac{1}{3}$$

Or, alternatively, first note that $AZ = 4$ because $\triangle JAZ$ is a $3 - 4 - 5$ right triangle. Then finish with the second part of the theorem.

c. Using the third part of the theorem,

$$(YZ)^2 = (AY)(JY)$$

$$= 6 \cdot 8$$

$$= 48$$

$$YZ = \sqrt{48} = 4\sqrt{3} \approx 6.9$$

d. Using the second part of the theorem,

$$(AZ)^2 = (JA)(AY)$$

$$8^2 = JA \cdot 10$$

$$JA = \frac{64}{10} = 6.4$$

$$JY = JA + AY$$

$$= 6.4 + 10$$

$$= 16.4$$

e. Using the third part of the theorem,

$$(JZ)^2 = (JA)(JY)$$

$$8^2 = JA \cdot 12$$

$$JA = \frac{64}{12} = 5\frac{1}{3}$$

$$AY = JY - JA$$

$$= 12 - 5\frac{1}{3}$$

$$= 6\frac{2}{3}$$

16) The Pythagorean Theorem gives you QS:

$$(QS)^2 = (RQ)^2 + (RS)^2$$

$$= 5^2 + 10^2$$

$$= 125$$

$$QS = \sqrt{125} = 5\sqrt{5} \approx 11.2$$

You can now get QT with the third part of the Altitude-on-Hypotenuse Theorem —
$(RQ)^2 = (QT)(QS)$ — and then get RT with the Pythagorean Theorem. But you don't
have to do all that. You can get RT directly with a proportion from similar triangles:

$$\frac{\text{long leg}_{\triangle QRT}}{\text{long leg}_{\triangle QSR}} = \frac{\text{hypotenuse}_{\triangle QRT}}{\text{hypotenuse}_{\triangle QSR}}$$

$$\frac{RT}{RS} = \frac{RQ}{QS}$$

$$\frac{RT}{10} = \frac{5}{5\sqrt{5}}$$

$$RT = \frac{10}{\sqrt{5}} = 2\sqrt{5} \approx 4.5$$

(*17) Set FL equal to x, and then use the last part of the Altitude-on-Hypotenuse Theorem:

$$(FG)^2 = (FL)(FA)$$

$$(6\sqrt{5})^2 = (x)(x+3)$$

$$180 = x^2 + 3x$$

$$x^2 + 3x - 180 = 0$$

Finish by factoring — $(x+15)(x-12) = 0$ — or with the quadratic formula. (If you forgot the
quadratic formula, I don't want to hear about it. But you can refresh your memory right
here.) Here's what the quadratic formula looks like in action:

For an equation in the form $ax^2 + bx + c = 0$,

$$x = \frac{-b \pm \sqrt{b^2 - 4ac}}{2a}$$

$$= \frac{-3 \pm \sqrt{3^2 - 4(1)(-180)}}{2(1)}$$

$$= \frac{-3 \pm \sqrt{729}}{2}$$

$$= \frac{-3 \pm 27}{2}$$

$$= -15 \text{ or } 12$$

FL can't be negative, so FL is 12.

(18) Here are the answers:

a. Because $\overline{OI} \parallel \overline{GN}$, $\angle ROI \cong \angle RGN$ and $\angle RIO \cong \angle RNG$ by *if parallel lines are cut by a transversal,
then corresponding angles are congruent.* Thus, $\triangle RIO \sim \triangle RNG$ by AA. (You can also use
$\angle R \cong \angle R$ for one of the two pairs of congruent angles.)

Keep your eyes peeled for parallel lines. Whenever you see parallel lines in a problem
involving two or more triangles, the odds are good that some of the triangles are similar
(or maybe congruent).

TIP

b. The Side-Splitter Theorem gives you *IN*:

$$\frac{RO}{OG} = \frac{RI}{IN}$$

$$\frac{4}{12} = \frac{6}{IN}$$

$$4 \cdot IN = 72$$

$$IN = 18$$

For *GN*, did you fall for my trap? *GN* is not 12, though it sure looks like it should be. The ratio of *OI* : *GN* doesn't equal the 4 : 12 ratio of *RO* : *OG*. Remember, the Side-Splitter Theorem doesn't work for the parallel sides. To get *GN*, you have to use the proportional sides of similar triangles *RIO* and *RNG* (*RG* = *RO* + *OG*, or 4 + 12):

$$\frac{\text{right side}_{\triangle RNG}}{\text{right side}_{\triangle RIO}} = \frac{\text{base}_{\triangle RNG}}{\text{base}_{\triangle RIO}}$$

$$\frac{GN}{OI} = \frac{RG}{RO}$$

$$\frac{GN}{4} = \frac{16}{4}$$

$$GN = 16$$

(*19) Along Adams, the distance from First to Third equals the distance from Third to Fifth. That's a 1 : 1 ratio. According to the transversals theorem, that 1 : 1 ratio must also hold for Washington and Jefferson.

The whole trip on Washington is half a mile, so you'd have to go 0.25 miles from First to Third and from Third to Fifth. Because it's 0.1 miles from Third to Fourth, Washington runs 0.25 − 0.1, or 0.15 miles, from Fourth to Fifth. The whole trip on Jefferson is 0.8 miles, so each half trip is 0.4 miles. Subtracting the 0.25 miles along First to Second from 0.4 gives you 0.15 miles for the distance from Second to Third.

Now use the distances along Washington from Third to Fourth (0.1 miles) and from Fourth to Fifth (0.15 miles) to get the corresponding distances along Adams and Jefferson. Along Washington, you have a ratio of 0.1 : 0.15, which equals 10 : 15, or 2 : 3. Using that ratio on Adams gives you

$$2x + 3x = \frac{3}{8}$$

$$5x = \frac{3}{8}$$

$$x = \frac{3}{40}$$

So, the distance along Adams from Third to Fourth ($2x$) is $2\left(\frac{3}{40}\right)$, or $\frac{6}{40}$, or 0.15 miles (that's three times in a row for 0.15 — what a bizarre coincidence! — I didn't plan it that way). From Fourth to Fifth ($3x$) is $3\left(\frac{3}{40}\right)$, or $\frac{9}{40}$, or 0.225 miles.

The calculation works the same along Jefferson, so

$$2x + 3x = 0.4$$

$$5x = 0.4$$

$$x = 0.08$$

So, the distance along Jefferson is 2(0.08), or 0.16 miles, from Third to Fourth, and 3(0.08), or 0.24 miles, from Fourth to Fifth.

Finally, use the same method with the distances along Jefferson from First to Second (0.25 miles) and from Second to Third (0.15 miles) to get the corresponding distances along Washington and Adams. The ratio along Jefferson is 0.25 : 0.15, which equals 5 : 3. So, for Washington, you have

$$5x + 3x = \frac{1}{4} \text{ mile}$$

$$8x = \frac{1}{4} \text{ mile}$$

$$x = \frac{1}{32} \text{ mile}$$

Thus, it's $5\left(\frac{1}{32}\right)$ or $\frac{5}{32}$ miles from First to Second and $3\left(\frac{1}{32}\right)$ or $\frac{3}{32}$ miles from Second to Third. For Adams, you have

$$5x + 3x = \frac{3}{8} \text{ miles}$$

$$x = \frac{3}{64} \text{ miles}$$

So, the distance along Adams is $\frac{15}{64}$ miles from First to Second and $\frac{9}{64}$ miles from Second to Third. That's it. *Finito!*

Street	Washington	Adams	Jefferson
Fourth to Fifth	$\frac{3}{20}$ (0.15)	$\frac{9}{40}$ (0.225)	$\frac{6}{25}$ (0.24)
Third to Fourth	$\frac{1}{10}$ (0.1)	$\frac{3}{20}$ (0.15)	$\frac{4}{25}$ (0.16)
Second to Third	$\frac{3}{32}$ (0.09375)	$\frac{9}{64}$ (0.140625)	$\frac{3}{20}$ (0.15)
First to Second	$\frac{5}{32}$ (0.15625)	$\frac{15}{64}$ (0.234375)	$\frac{1}{4}$ (0.25)
Total (miles)	$\frac{1}{2}$ (0.5)	$\frac{3}{4}$ (0.75)	$\frac{4}{5}$ (0.8)

4

Circles, Proof and Non-Proof Problems

Chapter 9

Circular Reasoning (Including Proofs)

The circle is a paradox of sorts. In one sense, it's the simplest of all shapes, but at the same time, it's rich in complexity and difficult, advanced ideas. Mathematicians have been fascinated by its properties for well over 2000 years. For example, the ratio of a circle's circumference to its diameter — $\pi \approx 3.14$ — is one of the most important and often-used numbers in all of mathematics.

In this chapter, you study several circle properties by doing proofs. All the proofs here involve circles, but you also see many of the same ideas from earlier chapters like *if angles, then sides* and CPCTC (both in Chapter 5).

The Segments Within: Radii and Chords

In this section, you do proofs involving radii and chords (including diameters).

REMEMBER

» **Radius:** Nothing about a circle is more fundamental than its radius. A circle's *radius* — the distance from its center to a point on the circle — tells you its size. In addition to being a measure of distance, a radius is also a segment from a circle's center to a point on the circle.

» **Chord:** A segment that connects two points on a circle is called a *chord*.

» **Diameter:** A chord passing through a circle's center is a *diameter* of the circle. A circle's diameter, as I'm sure you know, is twice as long as its radius.

Here are five circle theorems for your mathematical pleasure (some are the converses of each other).

» **Radii size:** All radii of a circle are congruent (yet another *well-duh* theorem).

» **Perpendicularity and bisected chords:**

• If a radius is perpendicular to a chord, then it bisects the chord.

• If a radius bisects a chord (that's not a diameter), then it's perpendicular to the chord.

» **Distance and chord size:**

• If two chords of a circle are equidistant from the center of the circle, then they're congruent.

• If two chords of a circle are congruent, then they're equidistant from its center.

TIP

If you're looking for tips for completing circle proofs, you've come to the right place. Here are a few that you can put to use in this section:

1. **Draw additional radii on the figure.**

   You should draw radii to points where something else intersects or touches the circle, as opposed to just any old point on the circle.

2. **Open your eyes and notice all the radii — including new ones you've drawn — and mark them all congruent.**

   For some strange reason — despite the fact that *all radii are congruent* is one of the simplest of all theorems — it's very common for people to either fail to notice all the radii in a problem or fail to note that they're all congruent.

3. **Draw in the segment (part of a radius) that goes from the center of the circle to a chord and that's perpendicular to the chord (and which, according to the previous theorem, bisects the chord).**

The purpose of adding radii and partial radii is usually to create right triangles or isosceles triangles that you can then use to solve the problem.

Q. Given: Circle O

                    $\overline{XY} \cong \overline{ZY}$

Prove: $\overline{YO}$ bisects $\angle XYZ$

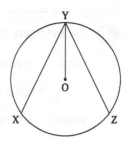

A.

Statements	Reasons
1) Circle O	1) Given.
2) Draw $\overline{OX}$ and $\overline{OZ}$	2) Two points determine a segment.
3) $\overline{OX} \cong \overline{OZ}$	3) All radii are congruent.
4) $\overline{XY} \cong \overline{ZY}$	4) Given.
5) $\overline{YO} \cong \overline{YO}$	5) Reflexive.
6) $\triangle YOX \cong \triangle YOZ$	6) SSS (3, 4, 5).
7) $\angle XYO \cong \angle ZYO$	7) CPCTC.
8) $\overline{YO}$ bisects $\angle XYZ$	8) Definition of bisect.

---

1   Given: Circle S

                    $\angle P \cong \angle R$

Prove: $\overline{PX} \cong \overline{RX}$

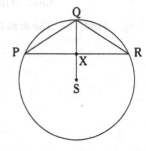

Statements	Reasons

2 Use the figure from problem 1, but this time with the following information.

Given: Circle S

  $\overline{QS} \perp \overline{PR}$

Prove: $\overrightarrow{QS}$ bisects $\angle PQR$

Statements	Reasons

3 Given: Isosceles trapezoid $ISTR$ with bases $\overline{ST}$ and $\overline{IR}$ is inscribed in circle Q
  Circle Q has a radius of 5

Find: The area of $ISTR$

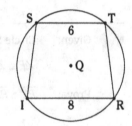

Given: Circle K

$\triangle FLT \cong \triangle AYT$

$\overline{KI} \perp \overline{FY}$ and $\overline{KE} \perp \overline{AL}$

Prove: *KITE* is a kite

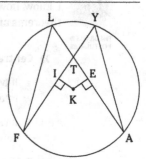

Hint: Use two of the five theorems from this section.

Statements	Reasons

Introducing Arcs and Central Angles

REMEMBER

Arcs, chords, and central angles are three peas in a pod.

>> **Arc:** An *arc,* as you may know, is simply a curved piece of a circle. Every chord cuts a circle into two arcs: a *minor arc* (the smaller piece) and a *major arc* (the larger), unless the chord is a diameter, in which case both arcs are semicircles.

>> **Central angle:** A *central angle* is an angle whose vertex is at the center of a circle. The two sides of a central angle are radii that hit the circle at the opposite ends of an arc or, as mathematicians say, the sides *intercept* an arc. The measure of an arc is the same as the degree measure of the central angle that intercepts it. (For more on arcs, chords, and angles that intercept an arc, see Chapter 10.)

REMEMBER

Congruent circles. Before I get into theorems, here's one more (somewhat unrelated) definition: *Congruent circles* are circles with congruent radii.

I know how much you love theorems, so here are six more. But don't sweat it — these six theorems are really just six variations on one simple idea about arcs, chords, and central angles.

>> **Central angles and arcs:**

* If two central angles of a circle (or of congruent circles) are congruent, then their intercepted arcs are congruent. (Short form: If central angles, then arcs.)

* If two arcs of a circle (or of congruent circles) are congruent, then the corresponding central angles are congruent. (Short form: If arcs, then central angles.)

>> **Central angles and chords:**

* If two central angles of a circle (or of congruent circles) are congruent, then the corresponding chords are congruent. (Short form: If central angles, then chords.)

* If two chords of a circle (or of congruent circles) are congruent, then the corresponding central angles are congruent. (Short form: If chords, then central angles.)

>> **Arcs and chords:**

* If two arcs of a circle (or of congruent circles) are congruent, then the corresponding chords are congruent. (Short form: If arcs, then chords.)

* If two chords of a circle (or of congruent circles) are congruent, then the corresponding arcs are congruent. (Short form: If chords, then arcs.)

Q. Given: Circle S

            $\angle P \cong \angle R$

    Prove: $\angle PSQ \cong \angle RSQ$

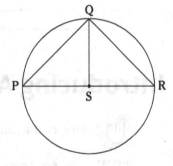

A.

Statements	Reasons
1) Circle S	1) Given.
2) $\angle P \cong \angle R$	2) Given.
3) $\overline{PQ} \cong \overline{RQ}$	3) If angles, then sides.
4) $\angle PSQ \cong \angle RSQ$	4) If chords, then central angles.

Short and sweet. Try doing this proof with congruent triangles instead. It should take you three extra steps.

5. **Given:** Circle Q

$\overline{AC} \cong \overline{RS}$

Prove: $\overline{AR} \cong \overline{CS}$

Warning: Note that the first four theorems in this section involve central angles (angles with a vertex at the center of a circle) and that, therefore, they *do not* apply to the angles in the figure to the right. (I cover angles like these in Chapter 10.)

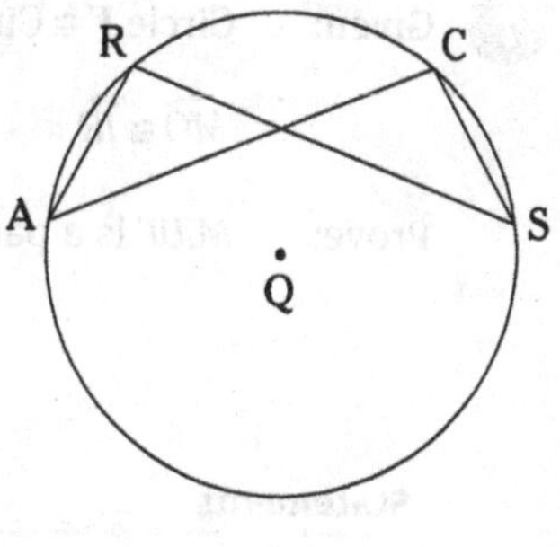

Statements	Reasons

6. **Given:** Circle Z

$\overline{BA} \cong \overline{CD}$

Prove: $\angle B \cong \angle C$

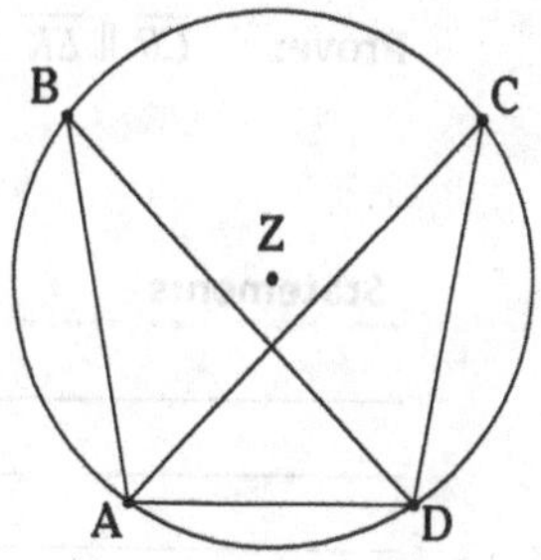

Statements	Reasons

7 Given: Circle $F \cong$ Circle U

   $\overparen{MO} \cong \overparen{IR}$

Prove: *MIUF* is a parallelogram

Statements	Reasons

8 Given: Circle $I \cong$ Circle L

   S is the midpoint of $\overline{CK}$

Prove: $\overline{CR} \parallel \overline{EK}$

Statements	Reasons

Touching on Radii and Tangents

One of the important ideas in this section is related to something you've seen since you were a little kid. Look at either wheel in the bicycle in Figure 9-1.

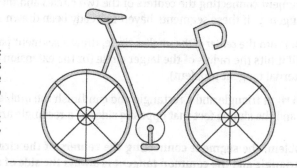

FIGURE 9-1:
Two wheels that are tangent to the ground — take a break from geometry and go for a ride.

A line is *tangent* to a circle if it touches the circle at a single point. The important point for this section is that the spoke that goes straight down from the hub in each wheel is *perpendicular* to the ground. Geometrically speaking, the bicycle wheels are, of course, circles, the spokes are radii, the single points where the wheels touch the ground are called *points of tangency*, and the ground is a *tangent* or *tangent line*.

THEOREMS & POSTULATES

Tangent and radius perpendicularity: A tangent line is perpendicular to the radius drawn to the point of tangency.

Don't forget this important fact! You already know how important it is to notice that all radii in a circle are congruent (and to sometimes draw in more radii). Now you can add this point about radii (and tangents) to your list of critical things to remember: The right angle at the point of tangency often becomes part of a right triangle.

Here's one more fact about tangents before I go through a couple of example problems.

THEOREMS & POSTULATES

Dunce Cap Theorem: Two tangent segments drawn to a circle from the same point are congruent. This is known (by me) as the Dunce Cap Theorem. See Figure 9-2.

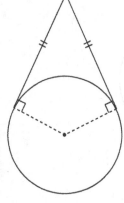

FIGURE 9-2:
Both sides of a dunce cap are the same length.

The first example and the first practice problem are called *common-tangent problems*, which means a single line is tangent to both circles. The example involves a common *external* tangent (the tangent lies on the same side of both circles). The practice problem involves a common *internal* tangent (the tangent line goes between the two circles). The solution method is the same for both.

1. **First:** Draw both the segment connecting the centers of the two circles and the two radii to the points of tangency (if these segments haven't already been drawn for you).

2. **This is the critical step:** From the *center of the smaller* circle, draw a segment parallel to the common tangent till it hits the radius of the larger circle (or the extension of the radius in a common internal tangent problem).

3. **Finish:** You now have a right triangle and a rectangle and can finish the problem with the Pythagorean Theorem and the simple fact that opposite sides of a rectangle are congruent.

REMEMBER

In a common-tangent problem, **the segment connecting the centers of the circles is** *always* **the hypotenuse of a right triangle and the common tangent is** *always* **the side of a rectangle.**

WARNING

In a common-tangent problem, **the segment connecting the centers of the circles is** *never* **one side of a right angle.**

EXAMPLE

Q. Given: $\overline{ET}$ is tangent to circle L and circle B with radii as shown

The distance between the centers of the circles is 25

Find: The length of the common tangent, $\overline{ET}$

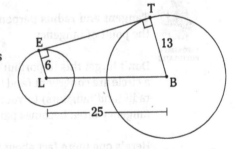

A. The segment connecting the centers of the two circles, as well as the two radii to the points of tangency (step 1 in solving common-tangent problems), are already drawn for you here. Draw the segment described in step 2 of the solution method (from the center of the smaller circle, parallel to the common tangent). See the following:

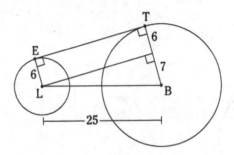

You can see that this new segment creates a rectangle and a right triangle. The opposite sides of a rectangle are congruent, so the radius of 6 on the left gives you the 6 on the right. The larger radius is 13, and 13 – 6 is 7, so you get 7 for the short leg of the right triangle. Its hypotenuse is 25, so you have a 7 – 24 – 25 right triangle, and thus the long leg is 24 (see info on Pythagorean triples in Chapter 3). Finally, the long leg of the triangle is also a side of the rectangle, which is congruent to the opposite side, $\overline{ET}$. ET is thus 24 as well.

9 Given: $\overline{IT}$ is a common internal tangent of circles S and L with radii as shown

A distance of 5 separates the circles

Find: IT (Get it?)

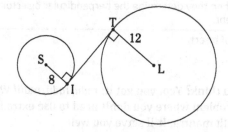

10 Given: Diagram as shown

$\overline{WR}$, $\overline{WL}$, $\overline{TL}$, and $\overline{TR}$ are tangent to circle O

Find: TR

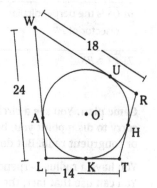

Solutions

(1) **Game plan:** You see a circle, so think *radii, radii, radii!* Draw in congruent radii $\overline{SP}$ and $\overline{SR}$. You also see an isosceles triangle, and, voilà, you can use *if angles, then sides* to get $\overline{PQ} \cong \overline{RQ}$. Then you can finish with the equidistance theorem (which first appears in Chapter 5).

Statements	Reasons
1) Circle S	1) Given.
2) Draw $\overline{SP}$ and $\overline{SR}$	2) Two points determine a segment.
3) $\overline{SP} \cong \overline{SR}$	3) All radii are congruent.
4) $\angle P \cong \angle R$	4) Given.
5) $\overline{PQ} \cong \overline{RQ}$	5) If angles, then sides.
6) $\overline{QS}$ is the perpendicular bisector of $\overline{PR}$	6) If two points are each equidistant from the endpoint *s* of a segment, then they determine the perpendicular bisector of that segment.
7) $\overline{PX} \cong \overline{RX}$	7) Definition of bisect.

(2) **Game plan:** You see a circle, so what should you think? Yep, you got it: *radii, radii, radii!* Well, sorry to disappoint you, but this is one circle problem where you don't need to use extra radii or congruent radii. But don't let up with the radii mantra. It'll serve you well.

You have a radius perpendicular to a chord, so the theorem tells you the chord is bisected. You can use that fact, the right angles, and the Reflexive Property to get the triangles congruent with SAS. Then you finish with — what else? — CPCTC (for more on congruent triangles, see Chapter 5).

Statements	Reasons
1) Circle S with $\overline{QS} \perp \overline{PR}$	1) Given.
2) $\overline{QS}$ bisects $\overline{PR}$	2) If a radius is perpendicular to a chord, then it bisects the chord.
3) $\overline{PX} \cong \overline{RX}$	3) Definition of bisect.
4) $\angle PXQ$ is a right angle $\angle RXQ$ is a right angle	4) Definition of perpendicular.
5) $\angle PXQ \cong \angle RXQ$	5) All right angles are congruent.
6) $\overline{QX} \cong \overline{QX}$	6) Reflexive.
7) $\triangle PQX \cong \triangle RQX$	7) SAS (3, 5, 6).
8) $\angle PQX \cong \angle RQX$	8) CPCTC.
9) $\overline{QS}$ bisects $\angle PQR$	9) Definition of bisect.

③ **Game plan:** *Radii, radii, radii, radii!* That's right — draw in four of 'em to *I, S, T,* and *R* (actually, you need only two, but it's probably easier to see how the problem works if you draw all four of them; and in any event, when it comes to drawing radii, too many is better than too few). Now you have two isosceles triangles, △*IQR* and △*SQT* (plus the two on the sides you'll use later). Next, draw in the altitudes of these triangles. See the following figure:

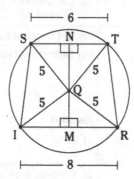

If a radius is perpendicular to a chord, it bisects the chord, so these altitudes bisect the bases of the triangles. That makes *IM* equal to 4 and *SN* equal to 3. And that makes both △*IMQ* and △*SNQ* 3–4–5 triangles (check out Chapter 4 for Pythagorean triples). Pretty sweet, eh? *QM* is thus 3 and *QN* is 4, so *NM* is 7, and that's the height of the trapezoid *ISTR*. You're all set to use the area formula (Chapter 7 provides info on calculating the area of quadrilaterals):

$$\text{Area}_{\text{Trap.}} = \frac{b_1 + b_2}{2} \cdot h$$
$$= \frac{8 + 6}{2} \cdot 7$$
$$= 7 \cdot 7$$
$$= 49 \text{ units}^2$$

Extra credit (well, maybe not exactly *credit*): Show that △*SQI* and △*TQR* are 45°–45°–90° triangles. You have two totally different ways of doing this. (This sure is a cool isosceles trapezoid, isn't it — the way it's made up of four congruent 3–4–5 triangles and two congruent 45°–45°–90° triangles.)

*④ **Abbreviated game plan:** Why would you be given congruent triangles? It's got to be so you can use CPCTC. You should notice the perpendicular segments drawn to the chords and think about how you could show the chords to be congruent. And you should, as always, also think about what you need at the end of the proof (namely, the two pairs of congruent sides that make a kite) and what you need to do to get there.

Statements	Reasons
1) △*FLT* ≅ △*AYT*	1) Given.
2) $\overline{FT} \cong \overline{AT}$	2) CPCTC.
3) $\overline{LT} \cong \overline{YT}$	3) CPCTC.
4) $\overline{FY} \cong \overline{AL}$	4) Segment addition.
5) $\overline{KI} \perp \overline{FY}$ $\overline{KE} \perp \overline{AL}$	5) Given.

6) $\overline{KI} \cong \overline{KE}$	6) If chords of a circle ($\overline{FY}$ and $\overline{AL}$) are congruent, then they are equidistant from its center.
7) $\overline{KI}$ bisects $\overline{FY}$ $\overline{KE}$ bisects $\overline{AL}$	7) If a radius is perpendicular to a chord, then it bisects the chord.
8) $\overline{IY} \cong \overline{EL}$	8) Like divisions. (Remember that? I knew you would. It's from Chapter 3.)
9) $\overline{IT} \cong \overline{ET}$	9) Segment subtraction (Statements 3 and 8).
10) *KITE* is a kite	10) Definition of kite.

5

Statements	Reasons
1) Circle Q $\overline{AC} \cong \overline{RS}$	1) Given.
2) $\overparen{AC} \cong \overparen{RS}$	2) If chords, then arcs.
3) $\overparen{AR} \cong \overparen{CS}$	3) Subtraction (subtracting $\overparen{RC}$ from both $\overparen{AC}$ and $\overparen{RS}$).
4) $\overline{AR} \cong \overline{CS}$	4) If arcs, then chords.

6

Statements	Reasons
1) Circle Z $\overline{BA} \cong \overline{CD}$	1) Given.
2) $\overparen{BA} \cong \overparen{CD}$	2) If chords, then arcs.
3) $\overparen{BD} \cong \overparen{CA}$	3) Addition (of $\overparen{AD}$).
4) $\overline{BD} \cong \overline{CA}$	4) If arcs, then chords.
5) $\overline{AD} \cong \overline{DA}$	5) Reflexive.
6) $\triangle ABD \cong \triangle DCA$	6) SSS (1, 4, 5).
7) $\angle B \cong \angle C$	7) CPCTC.

7

Statements	Reasons
1) Circle $F \cong$ Circle U $\overparen{MO} \cong \overparen{IR}$	1) Given.
2) $\angle MFO \cong \angle IUR$	2) If arcs, then central angles.
3) $\overline{MF} \parallel \overline{IU}$	3) If corresponding angles are congruent, then lines are parallel.
4) $\overline{MF} \cong \overline{IU}$	4) Congruent circles have congruent radii (definition of congruent circles).
5) *MIUF* is a parallelogram	5) If a quadrilateral has a pair of sides that are both parallel and congruent, then the quadrilateral is a parallelogram.

Statements	Reasons
1) Circle $I \cong$ Circle L S is the midpoint of $\overline{CK}$	1) Given.
2) $\overline{CS} \cong \overline{KS}$	2) Definition of midpoint.
3) $\overparen{CS} \cong \overparen{KS}$	3) If chords, then arcs.
4) $\angle CIR \cong \angle KLE$	4) Straight angles are congruent.
5) $\overparen{CSR} \cong \overparen{KSE}$	5) If central angles, then arcs.
6) $\overparen{SR} \cong \overparen{SE}$	6) Arc subtraction (subtracting congruent arcs from congruent arcs).
7) $\overline{SR} \cong \overline{SE}$	7) If arcs, then chords.
8) $\angle CSR \cong \angle KSE$	8) Vertical angles are congruent.
9) $\triangle CSR \cong \triangle KSE$	9) SAS (2, 8, 7).
10) $\angle C \cong \angle K$	10) CPCTC.
11) $\overline{CR} \parallel \overline{EK}$	11) If alternate interior angles are congruent, then lines are parallel.

Egad! I just saw a much better and easier way of doing this proof. Despite all my years of teaching geometry, I failed to follow my own advice about drawing in more radii. The preceding proof is a good illustration of some of the theorems in this section, but when it comes to proofs, the shorter the better. To wit —

Statements	Reasons
1) Circle $I \cong$ Circle L, S is the midpoint of $\overline{CK}$	1) Given.
2) $\overline{CS} \cong \overline{KS}$	2) Definition of midpoint.
3) Draw $\overline{IS}$ and $\overline{LS}$	3) Two points determine a segment.
4) $\overline{IS} \cong \overline{LS}$	4) Congruent circles have congruent radii.
5) $\overline{IC} \cong \overline{LK}$	5) Congruent circles have congruent radii.
6) $\triangle ICS \cong \triangle LKS$	6) SSS (2, 4, 5).
7) $\angle C \cong \angle K$	7) CPCTC.
8) $\overline{CR} \parallel \overline{EK}$	8) If alternate interior angles are congruent, then lines are parallel.

I guess you're never too old to learn.

(9) First, draw $\overline{SL}$. The middle portion of $\overline{SL}$ is the distance between the circles, and you were given that that distance is 5 units; therefore, SL is $8 + 5 + 12$, or 25. Next, from the center of the *smaller* circle, S, draw a segment parallel to $\overline{IT}$ till it hits the extension of radius $\overline{TL}$. Your diagram should now look like this:

The segment you just drew, $\overline{SR}$, creates a rectangle and a large right triangle. The rectangle has opposite sides of 8. The right triangle has a leg of $8 + 12 = 20$ and a hypotenuse of 25. That's in the $3 : 4 : 5$ family of triangles, so SR is 15, and that makes IT 15. That's it.

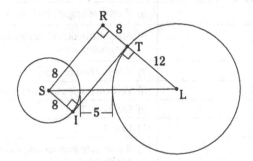

(10) Set HT equal to x, and walk around clockwise. Both sides of a dunce cap are equal, so KT is also x. KL is then $14 - x$, as is AL. Next, AW is $24 - (14 - x)$, which is $x + 10$. UW is also $x + 10$, and that makes UR equal to $18 - (x + 10)$, or $8 - x$. Finally, HR is also $8 - x$, and because HT is x, TR is $(8 - x) + x$, which is 8. That does it.

Chapter **10**

Scintillating Circle Formulas (No Proofs)

I f you're fully up-to-date on circle proofs, you're ready to move on to some handy circle formulas that help you calculate everything from area to arc length. In this chapter, you find the area and the perimeter of various sections of a circle. You also discover the relationships between angles whose vertices lie on, inside, and outside a circle and the intercepted arcs of these types of angles.

Pizzas, Slices, and Crusts: Finding Area and "Perimeter" of Circles, Sectors, and Segments

In this section, you work on problems involving the area and the circumference/perimeter of circles and parts of circles. (By the way, if the word *segment* in the heading is throwing you, you're not alone. *Segment* is the name of a particular section of a circle, which I show you in a minute. Don't ask me why the bozo who coined this term had to confuse matters by reusing a math term with another meaning.)

To start things off, here are two theorems — one about the area of a sector and a related theorem about the length of an arc. Don't worry: Both theorems are based on a very simple idea and are easier than they look. The mathematical examples following each of these theorems correspond to Figure 10-1.

» **Arc length:** The length of an arc (part of the circumference, like $\overparen{AB}$) is equal to the circumference of the circle (πd) times the fraction of the circle represented by the arc. $m\overparen{AB}$ is 30°, so

$$\text{Length}_{\overparen{AB}} = \left(\frac{m\overparen{AB}}{360} \right)\pi d$$

$$= \frac{30}{360} \cdot \pi \cdot 18$$

$$= \frac{1}{12} \cdot \pi \cdot 18$$

$$= 1.5\pi$$

» **Sector area:** The area of a sector (a wedge shape, like sector *AOB* in Figure 10-1) is equal to the area of the circle times the fraction of the circle represented by the sector.

$$\text{Area}_{\text{Sector } AOB} = \left(\frac{m\overparen{AB}}{360} \right)\pi r^2$$

$$= \frac{1}{12} \cdot \pi \cdot 9^2$$

$$= 6.75\pi \text{ units}^2$$

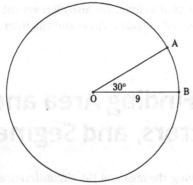

FIGURE 10-1:
A sector and
an arc that
make up
one-twelfth of
the circle.

TIP

Common sense suffices. These two theorems are so simple, you may want to do what I do — ignore them. I don't mean ignore the ideas; I mean you don't need to memorize the theorems. Your common sense should tell you that the length of $\overparen{AB}$ in this example is one-twelfth of the circumference of circle O because 30° goes into 360° twelve times. Likewise, the area of sector *AOB* is one-twelfth of the area of circle O. When common sense suffices, why clutter your mind with more formulas? Formulae, schmormulae.

EXAMPLE

Q. What's the area and perimeter of the shaded region? (This shape is the one that the aforementioned bozo decided to call a *segment*.)

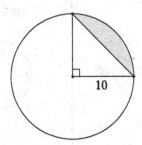

10

A. The area of the segment, as you can see, equals the area of the sector minus the area of the triangle. The sector measures 90°, so it's one-fourth of the circle. Thus,

$$\text{Area}_{\text{Sector}} = \frac{1}{4}\pi \cdot 10^2$$
$$= 25\pi \text{ units}^2$$

The area of the triangle is a no-brainer because its base and height are both radii:

$$\text{Area}_{\Delta} = \frac{1}{2}bh$$
$$= \frac{1}{2} \cdot 10 \cdot 10$$
$$= 50$$

Thus,

$$\text{Area}_{\text{Segment}} = \text{area}_{\text{sector}} - \text{area}_{\Delta}$$
$$= 25\pi - 50 \text{ units}^2$$

The perimeter equals the hypotenuse of the 45° – 45° – 90° triangle (which you can figure out in your head, right? — if not, turn to Chapter 4) plus one-fourth of the circle's circumference. To wit —

$$\text{Perimeter}_{\text{Segment}} = \text{hypotenuse} + \text{arc}$$
$$= 10\sqrt{2} + \frac{1}{4} \cdot 20\pi$$
$$= 10\sqrt{2} + 5\pi$$

1 Compute the area and perimeter of these shaded segments made from chords of length 6.

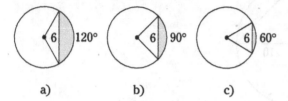

a)   b)   c)

2 Compute the shaded areas in the following figures. The inscribed polygons are regular, and each circle has a radius of 10.

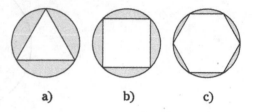

a)   b)   c)

3 Compute the shaded areas in the figures. The circumscribed polygons are regular, and each circle has a radius of 10.

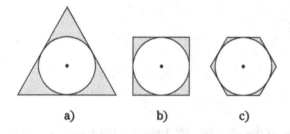

a)   b)   c)

*4 Compute the shaded area in the figure. The equilateral triangle is inscribed in a circle, and the three outer arcs are semicircles.

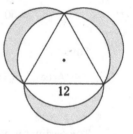

Angles, Circles, and Their Connections: The Angle-Arc Theorems and Formulas

Look at the circle in Figure 10-2 with $\overset{\frown}{AC}$ and $\angle ABC$.

Imagine that $\overline{BA}$ and $\overline{BC}$ are taut, elastic strings and that B is moveable. If you grab B and slide it around the edge of the circle (not crossing over A or C), $\angle B$ always stays the same size even though the distances from B to A and from B to C change. Isn't that cool? This and related ideas are the subjects of this section.

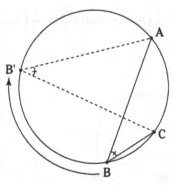

FIGURE 10-2:
For a given arc (like $\widehat{AC}$), no matter where you move the vertex of an inscribed angle, the angle measure doesn't change.

THEOREMS & POSTULATES

Angle on a circle: The measure of an inscribed angle (Figure 10-3a) or a tangent–chord angle (Figure 10-3b) is *one-half* of the measure of its intercepted arc.

For example, in the circles from Figure 10-3, $\angle Q = \frac{1}{2}\, m\widehat{PR}$ and $\angle Y = \frac{1}{2}\, m\widehat{XY}$.

a)

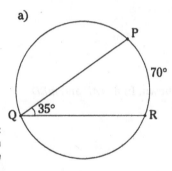

b)

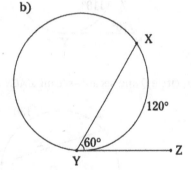

FIGURE 10-3:
Angles with a vertex *on* a circle.

THEOREMS & POSTULATES

Angle inside a circle: The measure of a chord-chord angle is *one-half* the *sum* of the measures of the arcs intercepted by the angle and its vertical angle.

For example, check out Figure 10-4: $\angle CED = \frac{1}{2}\left(m\widehat{AB} + m\widehat{CD} \right)$.

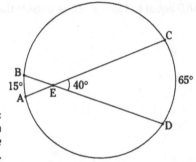

FIGURE 10-4:
An angle with a vertex *inside* a circle.

Angle outside a circle: The measure of a secant-secant angle (Figure 10-5a), a secant-tangent angle (Figure 10-5b), or a tangent-tangent angle (Figure 10-5c) is *one-half* the *difference* of the measures of the intercepted arcs.

For example, in the circles in Figure 10-5, $\angle C = \frac{1}{2}\left(m\widehat{AE} - m\widehat{BD}\right)$, $\angle R = \frac{1}{2}\left(m\widehat{PS} - m\widehat{QS}\right)$, and $\angle X = \frac{1}{2}\left(m\widehat{WZY} - m\widehat{WY}\right)$.

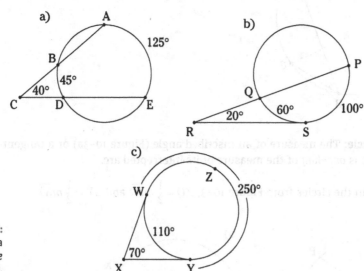

FIGURE 10-5:
Angles with a
vertex *outside*
a circle.

EXAMPLE

Q. Given circle Q and secant-secant $\angle ACE$ as shown, find $m\widehat{AE}$ and $m\widehat{BD}$.

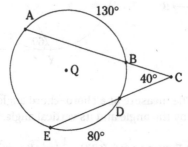

A. Because the measures of the four arcs $\widehat{AE}$, $\widehat{AB}$, $\widehat{BD}$, and $\widehat{ED}$ must add up to 360°, and because $m\widehat{AB}$ and $m\widehat{ED}$ add up to 210°, $m\widehat{AE}$ and $m\widehat{BD}$ must add up to 360° − 210°, or 150°. Now, set $m\widehat{AE}$ equal to x. That makes $m\widehat{BD}$ equal to 150 − x. $\angle C$ is *outside* the circle, so it equals half the *difference* of the arcs:

$$\angle C = \frac{1}{2}\left(m\widehat{AE} - m\widehat{BD}\right)$$

$$40 = \frac{1}{2}[x - (150 - x)]$$

$$80 = 2x - 150$$

$$230 = 2x$$

$$x = 115$$

Thus, $m\widehat{AE}$ is 115° and $m\widehat{BD}$ is 150° − 115°, or 35°.

5 Given the circle and angles as shown, find the measures of angles 1, 2, and 3.

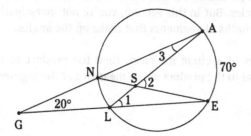

6 Given: Diagram as shown, with $\overline{AE}$ tangent to circle Q

Find: $\angle B$

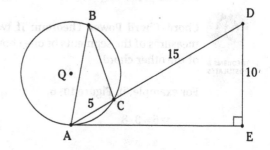

*7 Given: Circle C with a radius of 5

Find: XZ

Hint: You need to use one of the quadrilateral area formulas.

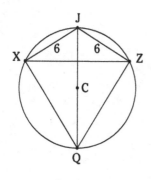

8 Given: Circle Q

$\angle MTE = 80°$

$\angle AMT = 70°$

Find: $\angle R$

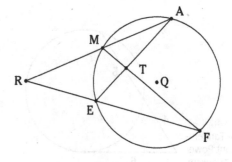

The Power Theorems That Be

The figures here look a lot like those from the preceding section because both sections involve angles drawn on the inside and outside of circles. But in this section, you're not investigating the size of the angles; you're looking at the lengths of segments that make up the angles.

THEOREMS & POSTULATES

Chord-Chord Power Theorem: If two chords of a circle intersect, then the product of the measures of the segments of one chord is equal to the product of the measures of the segments of the other chord.

For example, in Figure 10-6,

$$4 \cdot 6 = 3 \cdot 8$$

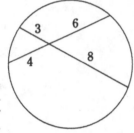

FIGURE 10-6: The Chord-Chord Power Theorem.

THEOREMS & POSTULATES

Tangent-Secant Power Theorem: If a tangent and a secant are drawn from an external point to a circle, then the square of the measure of the tangent is equal to the product of the measures of the secant's external part and the entire secant.

In Figure 10-7,

$$6^2 = 4 \cdot 9$$

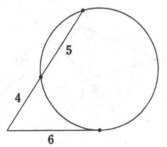

FIGURE 10-7: The Tangent-Secant Power Theorem.

THEOREMS & POSTULATES

Secant-Secant Power Theorem: If two secants are drawn from an external point to a circle, then the product of the measures of one secant's external part and that entire secant is equal to the product of the measures of the other secant's external part and that entire secant.

For instance, in Figure 10-8,

$$3 \cdot 12 = 4 \cdot 9$$

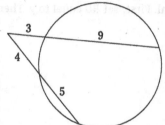

FIGURE 10-8:
The Secant-
Secant Power
Theorem.

TIP

All three of these theorems use the same mathematical idea. On both sides of each equation is a product of two lengths (or one length squared). Each of the four lengths is the distance from the vertex of an angle to the edge of the circle. Therefore, you can think of all three theorems like this:

$$(\text{vertex to circle}) \cdot (\text{vertex to circle}) = (\text{vertex to circle}) \cdot (\text{vertex to circle})$$

Pretty nifty, eh? It's because of ideas like this that they pay me the big bucks.

Q. Given: Diagram as shown

EXAMPLE

Circle Q has a radius of 7

Find: AB and RD

Note: It's difficult to see, but $\overline{DE}$ is a chord of circle Q; $\overline{RE}$ is not tangent to the circle at D or E.

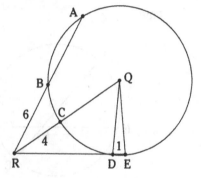

A. First, extend radius $\overline{CQ}$ into a diameter that hits the opposite side of the circle at a point I call X. The diameter has a length of 14. Now use the Secant–Secant Power Theorem to get AB:

$$RB \cdot RA = RC \cdot RX$$
$$6 \cdot RA = 4 \cdot 18$$
$$6 \cdot RA = 72$$
$$RA = 12$$

AB is RA minus RB, so AB is 6.

Finding *RD* is a bit trickier, but it's no big deal. First, set *RD* equal to *y*. Then

$$RD \cdot RE = RC \cdot RX$$
$$y(y+1) = 4 \cdot 18$$
$$y^2 + y = 72$$
$$y^2 + y - 72 = 0$$
$$(y+9)(y-8) = 0$$
$$y = -9 \text{ or } 8$$

You can reject −9, so *RD* has to be 8.

---

9   Given: Circle *O* has a radius of $5x + 1$

$$OR = 3x - 1$$
$$NR = 2x + 8$$
$$RY = 2x + 4$$

Find: *x*

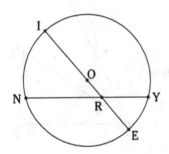

10   Given: Diagram as shown

$\overline{EA}$ is tangent to circle *Q*, which has a radius of 1

Find: *EA*, *EB*, and the area of $\triangle EQD$

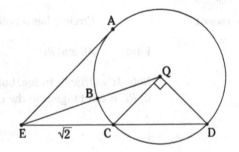

Solutions

(1) To tackle these problems, you may need to review some triangle formulas from Chapter 4.

    a. You need the radius. Draw the altitude of the triangle to the base of 6. The altitude bisects the 120° vertex angle and thus creates two 30° – 60° – 90° triangles, each with a long leg of 3. See the following figure:

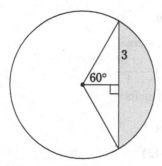

The short leg is $\dfrac{3}{\sqrt{3}}$, or $\sqrt{3}$, and the hypotenuse, therefore, is $2\sqrt{3}$ — that's the radius. You're all set:

$$\text{Area}_{\text{Segment}} = \text{area}_{\text{Sector}} - \text{area}_{\Delta}$$
$$= \frac{120°}{360°} \cdot \pi \left(2\sqrt{3}\right)^2 - \frac{1}{2}(6)\left(\sqrt{3}\right)$$
$$= \frac{1}{3}\pi \cdot 12 - 3\sqrt{3}$$
$$= 4\pi - 3\sqrt{3} \ \text{units}^2$$

You already have the base of the triangle, so the perimeter is a snap:

$$\text{Perimeter} = 6 + 120° \ \text{arc}$$
$$= 6 + \frac{1}{3}\pi d$$
$$= 6 + \frac{1}{3}\pi \left(4\sqrt{3}\right)$$
$$= 6 + \frac{4\pi\sqrt{3}}{3}$$

    b. You have a 45° – 45° – 90° triangle with a hypotenuse of 6, so the leg (which is the radius) is $\dfrac{6}{\sqrt{2}}$, or $3\sqrt{2}$. You're ready to go:

$$\text{Area}_{\text{Segment}} = \text{area}_{\text{Sector}} - \text{area}_{\Delta}$$
$$= \frac{90°}{360°}\pi \left(3\sqrt{2}\right)^2 - \frac{1}{2} \cdot \text{leg} \cdot \text{leg}$$
$$= \frac{1}{4}\pi \cdot 18 - \frac{1}{2}\left(3\sqrt{2}\right)\left(3\sqrt{2}\right)$$
$$= 4.5\pi - 9 \ \text{units}^2$$

$$\text{Perimeter} = 6 + 90° \text{ arc}$$

$$= 6 + \frac{1}{4}\pi d$$

$$= 6 + \frac{1}{4}\pi\left(6\sqrt{2}\right)$$

$$= 6 + \frac{3\pi\sqrt{2}}{2}$$

c. The easiest of the three. You have an equilateral triangle, so the radius is 6 and the sector takes up $\frac{1}{6}$ of the circle:

$$\text{Area}_{\text{Segment}} = \text{area}_{\text{Sector}} - \text{area}_{\Delta}$$

$$= \frac{1}{6}\left(\pi \cdot 6^2\right) - \frac{6^2\sqrt{3}}{4}$$

$$= 6\pi - 9\sqrt{3} \text{ units}^2$$

$$\text{Perimeter} = 6 + 60° \text{ arc}$$

$$= 6 + \frac{1}{6}\left(\pi \cdot 12\right)$$

$$= 6 + 2\pi$$

(2) Here's how this problem plays out.

a. Draw the apothem straight down to the base of the triangle (an apothem goes from the center of a regular polygon to the midpoint of a side — see Chapter 7). Then draw the circle's radius to one of the triangle's lower vertices. You now have a 30°–60°–90° triangle. The hypotenuse is the radius, so that's 10, and the apothem is the short leg, so that's 5. The long leg is, therefore, $5\sqrt{3}$, and because that's half the base of the triangle, the base is $10\sqrt{3}$ (see Chapter 4 for more on 30°–60°–90° triangles). You know all you need to solve the problem:

$$\text{Shaded area} = \text{circle} - \text{equilateral triangle}$$

$$= \pi r^2 - \frac{s^2\sqrt{3}}{4}$$

$$= \pi \cdot 10^2 - \frac{\left(10\sqrt{3}\right)^2\sqrt{3}}{4}$$

$$= 100\pi - 75\sqrt{3} \text{ units}^2$$

b. The radius is 10, and that's half of the square's diagonal, right? A square's a kite, so use the kite formula (Chapter 7):

$$\text{Shaded area} = \text{circle} - \text{square}$$

$$= \pi r^2 - \frac{1}{2}(d_1)(d_2)$$

$$= \pi \cdot 10^2 - \frac{1}{2}(20)(20)$$

$$= 100\pi - 200 \text{ units}^2$$

c. A regular hexagon is made up of six equilateral triangles (which you can read all about in Chapter 7). The radius here is 10, so the legs of the six triangles are also 10.

Shaded area = circle − hexagon

$$= \pi r^2 - 6\left(\text{area}_{\text{equil. } \triangle}\right)$$

$$= \pi \cdot 10^2 - 6\left(\frac{s^2\sqrt{3}}{4}\right)$$

$$= 100\pi - 6\left(\frac{10^2\sqrt{3}}{4}\right)$$

$$= 100\pi - 150\sqrt{3} \text{ units}^2$$

(3) Check out these solutions.

a. Because this problem is so similar to problem 2a, I'll cut to the chase (though note that in this problem, in contrast to 2a, the radius is now the short leg of a 30°–60°–90° triangle).

Shaded area = triangle − circle

$$= \frac{s^2\sqrt{3}}{4} - \pi \cdot 10^2$$

$$= \frac{\left(20\sqrt{3}\right)^2 \sqrt{3}}{4} - 100\pi$$

$$= 300\sqrt{3} - 100\pi \text{ units}^2$$

b. This one should be a no-brainer:

Shaded area = square − circle

$$= 20^2 - \pi \cdot 10^2$$

$$= 400 - 100\pi \text{ units}^2$$

c. Draw the apothem straight down, and draw a radius of the *hexagon* to one of its lower vertices. You have yet another 30°–60°–90° triangle. The apothem is the circle's radius, so it's 10. That's the long leg of the 30°–60°–90° triangle, so the short leg is $\frac{10}{\sqrt{3}}$. The short leg is half the length of one of the hexagon's sides, so those sides are $\frac{20}{\sqrt{3}}$, and multiplying that by 6 gives you the hexagon's perimeter: $\frac{120}{\sqrt{3}}$, or $40\sqrt{3}$. Use the formula for the area of a regular polygon (Chapter 6) for the hexagon.

Shaded area = hexagon − circle

$$= \frac{1}{2}pa - 100\pi$$

$$= \frac{1}{2}\left(40\sqrt{3}\right)(10) - 100\pi$$

$$= 200\sqrt{3} - 100\pi \text{ units}^2$$

(*4) The shaded area is everything minus the circle. And *everything* is the equilateral triangle plus the three semicircles. Thus,

Shaded area = triangle + 3 semicircles − circle

The triangle has sides of 12, and the radius (r_1) of each semicircle is 6, so you have everything you need to finish except for the radius (r_2) of the circle. You should be an expert at figuring this out by now. Draw the apothem straight down and draw a radius to one of the triangle's lower vertices. That gives you a 30°–60°–90° triangle. You can take it from there. You should get a radius (hypotenuse) of $4\sqrt{3}$.

$$\text{Shaded area} = \frac{s^2\sqrt{3}}{4} + 3\left(\frac{1}{2}\pi(r_1)^2\right) - \pi(r_2)^2$$

$$= \frac{12^2\sqrt{3}}{4} + 3\left(\frac{1}{2}\pi \cdot 6^2\right) - \pi\left(4\sqrt{3}\right)^2$$

$$= 36\sqrt{3} + 54\pi - 48\pi$$

$$= 36\sqrt{3} + 6\pi \text{ units}^2$$

Piece o' cake.

⑤ Here's how to find the angle measures.

$$\angle 1 = \frac{1}{2}m\widehat{AE}$$

$$= 35°$$

One down, two to go. You have a few ways to finish from this point. Here's one of them: ∠GLA is the supplement of ∠1, so it's 145°. Then, because the angles of △GLA have to add up to 180°, ∠3 is 15°. Two down, one to go.

$m\widehat{NL}$ is twice ∠3, so it's 30°. And so

$$\angle 2 = \frac{1}{2}\left(m\widehat{NL} + m\widehat{AE}\right)$$

$$= \frac{1}{2}(30 + 70)$$

$$= 50°$$

⑥ Right △ADE has a hypotenuse of 20 and a leg of 10. That makes it a 30°–60°–90° triangle (see Chapter 4 for more on special right triangles). You find that ∠CAE is thus 30°. The measure of tangent-chord ∠CAE is half of its intercepted arc, $\widehat{AC}$, so $m\widehat{AC}$ is 60°. Finally, the measure of inscribed ∠B is also half of $m\widehat{AC}$, so that makes it 30°.

I have a feeling that this problem may have been trickier than this short solution suggests.

⑦* ∠JXQ and ∠JZQ both intercept half the circle (a 180° arc), so each angle measures half of 180° — that's 90° of course — giving you right ∠JXQ and right ∠JZQ. The diameter is 10 — that's the hypotenuse — so legs $\overline{XQ}$ and $\overline{ZQ}$ are both 8 (the triangles are in the 3:4:5 family; see Chapter 4). Therefore, you have a 6–6–8–8 kite.

The area of the kite is twice the area of △JXQ, which is $\frac{1}{2} \cdot 6 \cdot 8$, or 24. So the area of the kite is 48 units². Now, finish with the kite area formula (from Chapter 7) and solve for the length of diagonal $\overline{XZ}$:

$$\text{Area}_{\text{Kite}} = \frac{1}{2}d_1 d_2$$

$$48 = \frac{1}{2}(10)(XZ)$$

$$48 = 5(XZ)$$

$$XZ = \frac{48}{5}, \text{ or } 9.6$$

Note: After you get the $6-8-10$ right triangles, you could also finish with the Altitude-on-Hypotenuse Theorem (see Chapter 7, which discusses similar triangles).

(8) $\angle AMF$ is inscribed in the circle, so $m\widehat{AF}$ is twice $\angle AMF$; thus, $m\widehat{AF}$ is 140°. Then to get $\angle R$ you just need the measure of $\widehat{ME}$:

$$\angle MTE = \frac{1}{2}\left(m\widehat{ME} + m\widehat{AF}\right)$$

$$80 = \frac{1}{2}\left(m\widehat{ME} + 140\right)$$

$$80 = \frac{1}{2}m\widehat{ME} + 70$$

$$m\widehat{ME} = 20°$$

And then

$$\angle R = \frac{1}{2}\left(m\widehat{AF} - m\widehat{ME}\right)$$

$$= \frac{1}{2}(140 - 20)$$

$$= 60°$$

(9) You want to use the Chord-Chord Power Theorem:

$$(\text{vertex to circle}) \cdot (\text{vertex to circle}) = (\text{vertex to circle}) \cdot (\text{vertex to circle})$$

For that, you need expressions in x for the lengths of the four segments. You have two of them: NR and RY. To get IR, you add $3x - 1$ to the radius of $5x + 1$. That's $8x$. And RE is the radius minus $3x - 1$. That's $2x + 2$. Now you have what you need to use the theorem:

$$(2x + 8)(2x + 4) = (8x)(2x + 2)$$

$$4x^2 + 8x + 16x + 32 = 16x^2 + 16x$$

$$-12x^2 + 8x + 32 = 0 \qquad (\text{Now divide both sides by } -4)$$

$$3x^2 - 2x - 8 = 0$$

$$(3x + 4)(x - 2) = 0$$

$$x = -\frac{4}{3} \text{ or } 2$$

You can reject $-\frac{4}{3}$, so x is 2.

(10) $\triangle CQD$ is an isosceles right triangle, which makes it a $45° - 45° - 90°$ triangle (see Chapter 4 for more information). The legs are 1, so the hypotenuse, $\overline{CD}$, is $\sqrt{2}$ units long. Now do the last problem first. To get the area of $\triangle EQD$, you need its height. So, draw the altitude of $\triangle CQD$ to base $\overline{CD}$. (This altitude of $\triangle CQD$ is also the altitude of $\triangle EQD$.) This altitude cuts $\triangle CQD$ into

two smaller 45°–45°–90° triangles. Each has a hypotenuse of 1 (the radius) and thus legs (including the altitude) of $\frac{1}{\sqrt{2}}$. The rest is child's play:

$$\text{Area}_{\triangle EQD} = \frac{1}{2}bh$$
$$= \frac{1}{2}\left(2\sqrt{2}\right)\left(\frac{1}{\sqrt{2}}\right)$$
$$= 1 \text{ unit}^2$$

To get EA, use the Tangent–Secant Power Theorem:

$$(\text{vertex to circle}) \cdot (\text{vertex to circle}) = (\text{vertex to circle}) \cdot (\text{vertex to circle})$$
$$(EA)(EA) = (EC)(ED)$$
$$(EA)^2 = \left(\sqrt{2}\right)\left(2\sqrt{2}\right)$$
$$(EA)^2 = 4$$
$$EA = 2$$

You use the Tangent–Secant Power Theorem again to get EB. First, set EB equal to x. Don't forget that you always go from vertex to *circle* (not from vertex to center of circle), so you need to use the whole diameter with a length of 2 that goes from B through Q to the other side of the circle:

$$(\text{vertex to circle}) \cdot (\text{vertex to circle}) = (\text{vertex to circle}) \cdot (\text{vertex to circle})$$
$$(EB)(EB + \text{diameter}) = (EA)(EA)$$
$$(x)(x+2) = (2)(2)$$
$$x^2 + 2x - 4 = 0$$

Finish up with the quadratic formula (which I first use in Chapter 4):

$$x = \frac{-2 \pm \sqrt{2^2 - 4(1)(-4)}}{2}$$
$$= \frac{-2 \pm \sqrt{20}}{2}$$
$$= \frac{-2 \pm 2\sqrt{5}}{2}$$
$$= -1 \pm \sqrt{5}$$

Reject the negative answer, so EB is $-1 + \sqrt{5}$. Wasn't that fun?

3-D Geometry: Proof and Non-Proof Problems

Chapter **11**

2-D Stuff Standing Up (Including Proofs)

M any of the ideas in this chapter should be familiar to you: congruent triangles, CPCTC, parallel lines, quadrilaterals, and so on. (If they're not, take a look at Chapters 1 through 10.) What's new about Chapter 11 is that some of the lines, triangles, and quadrilaterals, instead of lying in a plane, are now standing up in three-dimensional space.

Lines Perpendicular to Planes: They're All Right

This section involves problems about lines that are perpendicular to planes. Lines like this can come in handy, because they create right angles that are just begging for you to use them in a proof.

REMEMBER

Remember to look for *all* the right angles in the following problems; doing so can make the proofs much easier.

» **Plane:** A *plane* is a flat, two-dimensional shape — you know, like a piece of paper — except that it's infinitely thin and it goes on forever in all directions.

» **Perpendicularity of Line to Plane:** A line is perpendicular to a plane if it's perpendicular to every line in the plane that passes through its foot. (A *foot* is the point where a line intersects a plane.)

If a line is perpendicular to two lines that lie in a plane and pass through its foot, then it is perpendicular to the plane.

In short, when writing proofs, you **use the definition to say** *if a line is perpendicular to a plane, then it's perpendicular to a line in the plane (that passes through its foot)*, and you **use the theorem to say** *if a line is perpendicular to two lines in a plane, then it's perpendicular to the plane.*

Q. Given: $\overleftrightarrow{AB} \perp k$

EXAMPLE

BEDC is a kite with $\overline{BE} \cong \overline{BC}$

Prove: $\triangle AED \cong \triangle ACD$

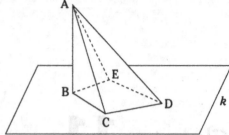

A.

Statements	Reasons
1) $\overline{AB} \perp k$	1) Given.
2) $\overline{AB} \perp \overline{BE}$ $\overline{AB} \perp \overline{BC}$	2) If a line is perpendicular to a plane, then it is perpendicular to every line in the plane that passes through its foot (definition of perpendicularity of a line to a plane).
3) $\angle ABE$ is a right angle $\angle ABC$ is a right angle	3) Definition of perpendicular.
4) $\angle ABE \cong \angle ABC$	4) All right angles are congruent.
5) $\overline{BE} \cong \overline{BC}$	5) Given.
6) $\overline{AB} \cong \overline{AB}$	6) Reflexive.
7) $\triangle ABE \cong \triangle ABC$	7) SAS (5, 4, 6).
8) $\overline{AE} \cong \overline{AC}$	8) CPCTC.
9) $\overline{ED} \cong \overline{CD}$	9) Property of a kite.
10) $\overline{AD} \cong \overline{AD}$	10) Reflexive.
11) $\triangle AED \cong \triangle ACD$	11) SSS (8, 9, 10).

1 Given: $\overline{BD} \perp p$

$\overrightarrow{AC}$ bisects $\angle BAD$

Prove: C is the midpoint of $\overline{BD}$

Statements	Reasons

2 Given: $\overline{VX} \perp q$

$\triangle VYZ$ is isosceles with base $\overline{YZ}$

Prove: $\triangle XYZ$ is isosceles

Statements	Reasons

3 Given: $\overline{ST} \perp r$

$\angle TVU \cong \angle TUV$

Prove: $\angle SVU \cong \angle SUV$

Statements	Reasons

*4 Given: Circle O in plane p

$\angle AOZ$ and $\angle BOZ$ are right angles

$\overset{\frown}{AB} \cong \overset{\frown}{BC}$

Prove: $\angle AZB \cong \angle CZB$

Statements	Reasons

Parallel, Perpendicular, and Intersecting Lines and Planes

In the preceding section, all the figures contain a single line perpendicular to a single plane. In this section, you move on to figures that involve multiple perpendicularity and/or multiple planes and parallel lines. But first take a look at the four ways to determine a plane.

**THEOREMS &
POSTULATES**

Determining a plane: Four different sets of geometric objects determine a plane:

>> Three noncollinear points

In plain English, this statement just means that if you have three points not on one line, then only one specific plane contains those points. The plane is *determined* by the three points because they show you exactly where this plane is.

>> A line and a point not on the line

>> Two intersecting lines

>> Two parallel lines

REMEMBER

Check out the following properties about perpendicularity and parallelism of lines and planes. For the most part, these are *well-duh* properties after you picture what the lines and planes would look like.

>> **Three parallel planes:** If two planes are parallel to the same plane, they're parallel to each other.

>> **Two parallel lines and a plane:**

● If two lines are perpendicular to the same plane, they're parallel to each other.

● If a plane is perpendicular to one of two parallel lines, it's perpendicular to the other.

>> **Two parallel planes and a line:**

● If two planes are perpendicular to the same line, they're parallel to each other.

● If a line is perpendicular to one of two parallel planes, it's perpendicular to the other.

And here's one more point before getting to the problems:

**THEOREMS &
POSTULATES**

Intersecting planes: If a plane intersects two parallel planes, the lines of intersection are parallel.

EXAMPLE

Q. Given: $\overline{GN} \perp p$

$\overline{LE} \perp p$

$\overline{GN} \cong \overline{LE}$

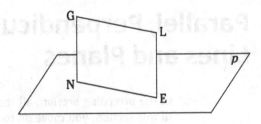

Prove: NGLE is a rectangle

A.

Statements	Reasons
1) $\overline{GN} \perp p$ $\overline{LE} \perp p$	1) Given.
2) $\overline{GN} \parallel \overline{LE}$	2) If two lines are perpendicular to the same plane, then they are parallel to each other.
3) $\overline{GN}$ and $\overline{LE}$ determine a plane, NGLE	3) Two parallel lines determine a plane. (You need this odd-looking step to ensure that NGLE is a planar quadrilateral. Otherwise, it could be a weird, bent, four-sided figure like a "rectangle" bent along one of its diagonals. It could also be a "rectangle" with a curvy surface. Can you picture these shapes?)
4) $\overline{GN} \cong \overline{LE}$	4) Given.
5) NGLE is a parallelogram	5) If a quadrilateral contains a pair of sides that are both parallel and congruent, then the quadrilateral is a parallelogram.
6) $\overline{GN} \perp \overline{NE}$	6) If a line is perpendicular to a plane, then it is perpendicular to every line in the plane that passes through its foot.
7) $\angle GNE$ is a right angle	7) Definition of perpendicular.
8) NGLE is a rectangle	8) A parallelogram with a right angle is a rectangle.

EXAMPLE

Q. Given: $p \parallel q$

$\overline{RT} \parallel \overline{SU}$

Prove: $\overline{RS} \cong \overline{TU}$

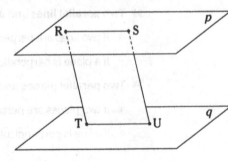

A.

Statements	Reasons
1) $p \parallel q$	1) Given.
2) $\overline{RT} \parallel \overline{SU}$	2) Given.
3) $\overline{RT}$ and $\overline{SU}$ determine a plane, RSUT	3) Two parallel lines determine a plane. (You need this step before you can use the theorem in Reason 4.)
4) $\overline{RS} \parallel \overline{TU}$	4) If a plane intersects two parallel planes, then the lines of intersection are parallel.
5) RSUT is a parallelogram	5) If both pairs of opposite sides of a quadrilateral are parallel, then the quadrilateral is a parallelogram (definition of parallelogram).
6) $\overline{RS} \cong \overline{TU}$	6) Opposite sides of a parallelogram are congruent.

5 Given: $s \parallel y$

$\triangle EIO$ is isosceles with base $\overline{EO}$

Prove: $\triangle AIU$ is isosceles

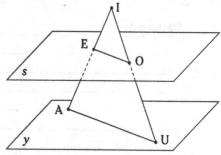

Statements	Reasons

6 Give this problem a go:

a. Given: $x \parallel y$

$\overline{MR} \parallel \overline{ED}$

$\overline{MR} \perp \overline{MD}$

Prove: $MRED$ is a rectangle

Yes or No: Is $\overline{MR} \perp y$?

Is $\overline{MR} \perp x$?

b. Given: $x \parallel y$

$\overline{MR} \parallel \overline{ED}$

$\overline{MR} \perp y$

Prove: $\overline{ED} \perp x$

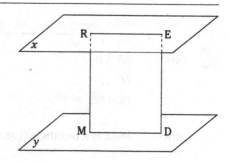

a.

Statements	Reasons

b.

Statements	Reasons

7 Given: $\overline{GN} \perp p$

$\overline{LE} \perp p$

$\overline{GL} \cong \overline{NE}$

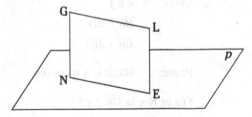

Prove: NGLE is a rectangle (paragraph proof)

(Note the similarity of this problem to the first example. The one minor difference here makes this problem *much* harder.)

Solutions

(1)

Statements	Reasons
1) $\overline{BD} \perp p$	1) Given.
2) $\overline{BD} \perp \overline{AC}$	2) If a line is perpendicular to a plane, then it's perpendicular to every line in the plane that passes through its foot.
3) $\angle BCA$ is a right angle $\angle DCA$ is a right angle	3) Definition of perpendicular.
4) $\angle BCA \cong \angle DCA$	4) All right angles are congruent.
5) $\overline{AC}$ bisects $\angle BAD$	5) Given.
6) $\angle BAC \cong \angle DAC$	6) Definition of bisect.
7) $\overline{AC} \cong \overline{AC}$	7) Reflexive.
8) $\triangle BAC \cong \triangle DAC$	8) ASA (4, 7, 6).
9) $\overline{BC} \cong \overline{DC}$	9) CPCTC.
10) C is the midpoint of $\overline{BD}$	10) Definition of midpoint.

(2)

Statements	Reasons
1) $\overline{VX} \perp q$	1) Given.
2) $\overline{VX} \perp \overline{WZ}$ $\overline{VX} \perp \overline{WY}$	2) If a line is perpendicular to a plane, then it's perpendicular to every line in the plane that passes through its foot.
3) $\angle VWZ$ is a right angle $\angle VWY$ is a right angle	3) Definition of perpendicular.
4) $\triangle VYZ$ is isosceles with base $\overline{YZ}$	4) Given.
5) $\overline{VZ} \cong \overline{VY}$	5) Definition of isosceles triangle.
6) $\overline{VW} \cong \overline{VW}$	6) Reflexive.
7) $\triangle VWZ \cong \triangle VWY$	7) HLR (5, 6, 3).
8) $\angle ZVW \cong \angle YVW$	8) CPCTC.
9) $\overline{VX} \cong \overline{VX}$	9) Reflexive.
10) $\triangle ZVX \cong \triangle YVX$	10) SAS (5, 8, 9).
11) $\overline{ZX} \cong \overline{YX}$	11) CPCTC.
12) $\triangle XYZ$ is isosceles	12) Definition of isosceles triangle.

3	Statements	Reasons
1) $\overline{ST} \perp r$	1) Given.
2) $\overline{ST} \perp \overline{TV}$ $\overline{ST} \perp \overline{TU}$	2) If a line is perpendicular to a plane, then it's perpendicular to every line in the plane that passes through its foot.
3) $\angle STV$ is a right angle $\angle STU$ is a right angle	3) Definition of perpendicular.
4) $\angle STV \cong \angle STU$	4) All right angles are congruent.
5) $\angle TVU \cong \angle TUV$	5) Given.
6) $\overline{TV} \cong \overline{TU}$	6) If angles, then sides.
7) $\overline{ST} \cong \overline{ST}$	7) Reflexive.
8) $\triangle STV \cong \triangle STU$	8) SAS (6, 4, 7).
9) $\overline{SV} \cong \overline{SU}$	9) CPCTC.
10) $\angle SVU \cong \angle SUV$	10) If sides, then angles.

*4	Statements	Reasons
1) Circle O in plane p $\angle AOZ$ and $\angle BOZ$ are right angles	1) Given.
2) $\overline{OZ} \perp p$	2) If a line is perpendicular to two lines that lie in a plane and pass through its foot, then it is perpendicular to the plane.
3) Draw radius $\overline{OC}$	3) Two points determine a segment.
4) $\overline{OZ} \perp \overline{OC}$	4) If a line is perpendicular to a plane, then it's perpendicular to every line in the plane that passes through its foot.
5) $\angle COZ$ is a right angle	5) Definition of perpendicular.
6) $\angle AOZ \cong \angle COZ$	6) All right angles are congruent.
7) $\overline{OA} \cong \overline{OC}$	7) All radii are congruent.
8) $\overline{ZO} \cong \overline{ZO}$	8) Reflexive.
9) $\triangle AOZ \cong \triangle COZ$	9) SAS (7, 6, 8).
10) $\overline{ZA} \cong \overline{ZC}$	10) CPCTC.
11) Draw chords $\overline{AB}$ and $\overline{BC}$	11) Two points determine a segment.
12) $\overparen{AB} \cong \overparen{BC}$	12) Given.
13) $\overline{AB} \cong \overline{BC}$	13) If arcs, then chords.
14) $\overline{ZB} \cong \overline{ZB}$	14) Reflexive.
15) $\triangle AZB \cong \triangle CZB$	15) SSS (10, 13, 14).
16) $\angle AZB \cong \angle CZB$	16) CPCTC.

To find out more about arcs and circles, read up on them in Chapter 9.

Statements	Reasons
1) $s \parallel y$	1) Given.
2) $\overline{IA}$ and $\overline{IU}$ determine a plane, $AEIOU$	2) Two intersecting lines determine a plane.
3) $\overline{EO} \parallel \overline{AU}$	3) If a plane intersects two parallel planes, then the lines of intersection are parallel.
4) $\triangle EIO$ is isosceles with base $\overline{EO}$	4) Given.
5) $\overline{IE} \cong \overline{IO}$	5) Definition of isosceles triangle.
6) $\angle IEO \cong \angle IOE$	6) If sides, then angles.
7) $\angle IEO \cong \angle IAU$ $\angle IOE \cong \angle IUA$	7) If lines are parallel, then corresponding angles are congruent.
8) $\angle IAU \cong \angle IUA$	8) Transitivity.
9) $\overline{IA} \cong \overline{IU}$	9) If angles, then sides.
10) $\triangle AIU$ is isosceles	10) Definition of isosceles triangle.

⑥ Here are the answers.

a.

Statements	Reasons
1) $x \parallel y$	1) Given.
2) $\overline{MR} \parallel \overline{ED}$	2) Given.
3) $\overline{MR}$ and $\overline{ED}$ determine plane $MRED$	3) Two parallel lines determine a plane.
4) $\overline{RE} \parallel \overline{MD}$	4) If a plane intersects two parallel planes, then the lines of intersection are parallel.
5) $MRED$ is a parallelogram	5) Definition of parallelogram.
6) $\overline{MR} \perp \overline{MD}$	6) Given.
7) $\angle RMD$ is a right angle	7) Definition of perpendicular.
8) $MRED$ is a rectangle	8) A parallelogram with a right angle is a rectangle.

Answer to the Yes or No question: $\overline{MR}$ may or may not be perpendicular to plane y. (Remember, to know that a line is perpendicular to a plane, you must know that it is perpendicular to *two* lines in the plane that pass through its foot, not just one line. It's possible that $MRED$ is slanting toward or away from you.) If $\overline{MR}$ is perpendicular to plane y, it's perpendicular to plane x as well. If it's not perpendicular to y, it's also not perpendicular to x.

b.

Statements	Reasons
1) $x \parallel y$ $\overline{MR} \parallel \overline{ED}$ $\overline{MR} \perp y$	1) Given.
2) $\overline{ED} \perp y$	2) If a plane is perpendicular to one of two parallel lines, then it is perpendicular to the other.
3) $\overline{ED} \perp x$	3) If a line is perpendicular to one of two parallel planes, then it is perpendicular to the other.

**7) This proof begins just like the first three lines of the example proof: $\overline{GN}$ and $\overline{LE}$ are perpendicular to p (given), $\overline{GN}$ is parallel to $\overline{LE}$ (the two lines are perpendicular to the same plane), and $\overline{GN}$ and $\overline{LE}$ determine plane $NGLE$.

Then you state that $\overline{LE} \perp \overline{NE}$, because if a line is perpendicular to a plane, the line is perpendicular to any line in the plane that passes through its foot. And then you have that $\angle E$ is a right angle (you can show that $\angle N$ is a right angle the same way, but it doesn't help). Okay, so now you have a quadrilateral with two parallel sides ($\overline{GN}$ and $\overline{LE}$) — call them *bases* — where the other two sides are congruent and where one base angle is a right angle:

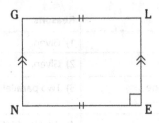

Now comes the tricky part.

Even though it's obvious that $NGLE$ must be a rectangle, I couldn't find a way to prove it with ordinary techniques. I drew in diagonals $\overline{GE}$ and $\overline{NL}$, and I tried to use things like *alternate interior angles are congruent* to get congruent triangles (see Chapter 6). I wanted to show that $\overline{GL} \parallel \overline{NE}$ or that $\angle L$ is a right angle. Nothing worked. (If anyone out there finds a way to finish this proof with ordinary methods, please let me know about it.)

Here's how I finished the proof. The only quadrilaterals with parallel bases in which the other two sides are congruent are parallelograms and isosceles trapezoids (see Chapter 7). Assume $NGLE$ is an isosceles trapezoid. Its base angles would therefore be congruent, and that would make $\angle L$ a right angle. But if $\angle E$ and $\angle L$ were right angles, then $NGLE$ would be a rectangle, which contradicts the assumption (because a rectangle is not an isosceles trapezoid). Therefore, $NGLE$ must be a parallelogram. This parallelogram has a right angle, and therefore it's a rectangle (one of the ways of proving that a parallelogram is a rectangle; see Chapter 6). Bingo. That's it.

IN THIS CHAPTER

» Finding the surface area and
volume of cylinders and prisms

» Calculating the area and volume of
cones and pyramids

» Having a ball with spheres

Chapter 12

Solid Geometry: Digging into Volume and Surface Area (No Proofs)

When working in flat, 2-D space previously in this book, I introduce lines and angles and then move on to planar shapes like triangles and parallelograms. Now I delve into more than just flat things. In this chapter, you can take a look at all kinds of new and fun figures in the next dimension — cylinders, prisms, cones, pyramids, and spheres (got your 3-D glasses handy?).

Starting with Flat-Top Figures

Flat-top figure is my nontechnical name for a cylinder or a prism. Both figures have — guess what? — a flat top. This flat top is called a *base*, and it's congruent to and parallel to the other base at the bottom of the figure. I group prisms (whose bases are polygons) and cylinders together because computing their volume basically works the same way; ditto for computing surface area:

REMEMBER

> » **Volume of flat-top figures.** The volume of a prism or cylinder is given by the following formula:
>
> $$\text{Vol}_{\text{Flat-Top}} = \text{area}_{\text{base}} \cdot \text{height}$$

» **Surface area of flat-top figures.** To find the surface area of a prism or a cylinder, use the following formula:

$$SA_{\text{Flat-Top}} = 2 \cdot \text{area}_{\text{base}} + \text{lateral area}_{\text{rectangle(s)}}$$

The *lateral area* (that's the area of the sides of the figure — namely, everything but the bases) of a right prism is made up of rectangles. The lateral area of a right cylinder is basically one rectangle rolled into a tube-shape — like one paper towel that rolls exactly once around a paper towel roll. The base of this rectangle (you know, its length) is thus the circumference of the cylinder.

Q. A cylinder with a volume of 125π units^3 has a height equal to its radius. Find its surface area.

EXAMPLE

A. First, use the volume formula:

$$\text{Vol}_{\text{Flat-Top}} = \text{area}_{\text{base}} \cdot \text{height}$$
$$125\pi = \pi r^2 \cdot h$$
$$125\pi = \pi r^3 \quad (\text{because } h = r)$$
$$r = 5 \quad (\text{and, therefore, } h = 5)$$

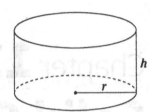

Now you can compute the surface area:

$$SA_{\text{Flat-Top}} = 2 \cdot \text{area}_{\text{base}} + \text{lateral area}_{\text{rectangle}}$$
$$= 2\pi r^2 + 2\pi r \cdot h$$
$$= 2\pi \cdot 5^2 + 2\pi \cdot 5 \cdot 5$$
$$= 50\pi + 50\pi$$
$$= 100\pi \text{ units}^2$$

---

1 Find the volume and surface area of this prism.

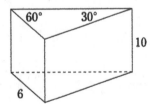

2 Find the volume and surface area of this prism, whose bases are equilateral triangles.

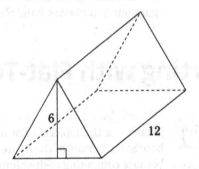

*3 Find the volume and surface area of a box (technically a prism) with a height of 2, a width of $2\sqrt{3}$, and a diagonal of 8.

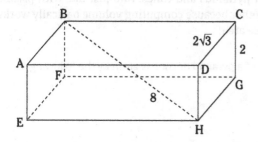

4 Answer the following:

a. What's the volume and surface area of a cylinder with height and diameter both equal to 4?

*b. An ant crawls along the outside of the cylinder from *A* to *C*. If the ant goes straight across the top to *B* and then straight down to *C*, it goes a distance of 8. Is there a shorter route? If so, what's the shortest possible route, and how long is it?

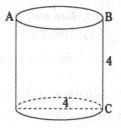

5 A cylinder has a diameter of 6 and a lateral area of 60π. Find its volume and surface area.

*6 A cylinder with a height of 6 has a surface area of 54π. Find its volume.

Sharpening Your Skills with Pointy-Top Figures

REMEMBER

Something tells me that you've already figured out that *pointy-top figures* are figures with pointy tops. This is my nontechnical name for pyramids and cones. And just like with prisms and cylinders, I group pyramids and cones together because computing volume basically works the same for both — as does computing surface area:

>> **Volume of pointy-top figures.** The volume of a pyramid or cone is given by the following formula:

$$Vol_{Pointy-Top} = \frac{1}{3} \, area_{base} \cdot height$$

>> **Surface area of pointy-top figures.** The following formula gives you the surface area of a pyramid or cone:

$$SA_{Pointy-Top} = area_{base} + lateral \, area_{triangle(s)}$$

The lateral area of a pyramid is made up of triangles whose areas work just like the area of any triangle: $\frac{1}{2} \, base \cdot height$. But note that the height of a triangle is perpendicular to its base, so you can't use the height of the pyramid for the height of one of its triangular faces. Instead, you use the *slant height*, which is just the ordinary height of the triangular face — if you look at the face like an ordinary flat, two-dimensional triangle. (The cursive letter ℓ is used to indicate slant height.)

Just like the lateral area of a cylinder is one rectangle rolled around into a tube-shape, the lateral area of a cone is one triangle (sort of — its bottom is curved) rolled around into a shape like a snow-cone cup. Its area works exactly like the area of one of the triangular faces of a pyramid, $\frac{1}{2} (base)(slant \, height)$, where the base of this "triangle" (just like the base of the lateral rectangle in a cylinder) equals the circumference of the cone.

EXAMPLE

Q. Find the volume and surface area of these similar cones (that's *similar* in the technical sense — see Chapter 8 for more on similarity), one of which has dimensions that are double the other. What do you notice about the answers?

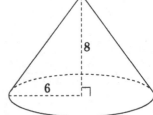

A. Doing volume first,

$$\text{Vol}_{\text{Small Cone}} = \frac{1}{3} \, \text{area}_{\text{base}} \cdot \text{height}$$

$$= \frac{1}{3} \pi r^2 h$$

$$= \frac{1}{3} \pi \cdot 3^2 \cdot 4$$

$$= 12\pi \text{ units}^3$$

$$\text{Vol}_{\text{Large Cone}} = \frac{1}{3} \pi r^2 h$$

$$= \frac{1}{3} \pi \cdot 6^2 \cdot 8$$

$$= 96\pi \text{ units}^3$$

The volume of the large cone is eight times the volume of the small one.

To find the cones' surface areas, you need their slant heights. If you use the height of a cone and one of its radii to form the legs of a right triangle, then the hypotenuse of the triangle is the cone's slant height. For the small cone in this problem, you have a $3-4-5$ right triangle, so the slant height is 5 (see Chapter 4 for more on Pythagorean triples). In the large cone, the slant height is 10. Now you can compute their surface areas:

$$\text{SA}_{\text{Small Cone}} = \text{area}_{\text{base}} + \text{lateral triangle}$$

$$\left(\frac{1}{2} \cdot \text{base} \cdot \text{height} \right)$$

$$\downarrow \qquad \searrow$$

$$= \pi r^2 + \frac{1}{2} (\text{circumference})(\text{slant height})$$

$$= \pi r^2 + \frac{1}{2} (2\pi r)(5)$$

$$= 9\pi + \frac{1}{2} (2\pi \cdot 3)(5)$$

$$= 9\pi + 15\pi$$

$$= 24\pi \text{ units}^2$$

$$\text{SA}_{\text{Large Cone}} = \pi r^2 + \frac{1}{2} (2\pi r)(\text{slant height})$$

$$= \pi \cdot 6^2 + \frac{1}{2} (2\pi \cdot 6)(10)$$

$$= 36\pi + 60\pi$$

$$= 96\pi \text{ units}^2$$

The large cone has four times the surface area of the small cone. So, the large cone, which is twice the size of the small cone, has eight $\left(2^3\right)$ times as much volume and four $\left(2^2\right)$ times as much surface area.

See the rule? Here it is:

The squaring and cubing rule for similar 3-D shapes. If you enlarge a 3-D figure by a factor of k, its surface area grows k^2 times and its volume grows k^3 times.

A good way to remember this rule is to note the connection between the rule and the fact that surface area is two-dimensional and is measured in $units^2$ and that volume is three-dimensional and is measured in $units^3$.

---

7 Find the volume and surface area of this rectangular, right pyramid.

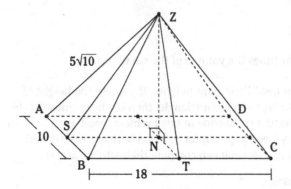

*8 Find the volume and surface area of a regular tetrahedron with edges of 6. (A regular *tetrahedron* is a pyramid with four equilateral triangle faces.)

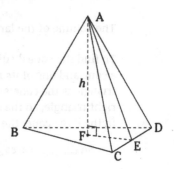

The circumference of the base of a cone is 16π, and the cone's height is 6. Find the cone's volume and surface area.

10 Try this one on for size:

*a. Find the volume of this double cone, which has a radius of 8 and a total height of 21.

b. If the lateral surface area of the left-side cone is 80π, what's the lateral surface area of the right-side cone?

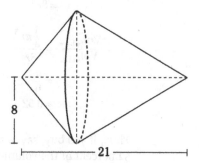

Rounding Out Your Understanding with Spheres

REMEMBER

I'm running short on space, so I better cut to the chase:

>> **Volume of a sphere.** The volume of a sphere is given by the following formula:

$$\text{Vol}_{\text{Sphere}} = \frac{4}{3}\pi r^3$$

>> **Surface area of a sphere.** Yada, yada, yada:

$$\text{SA}_{\text{Sphere}} = 4\pi r^2$$

Q. How do the volumes of a cube and an inscribed sphere compare? How do their surface areas compare?

EXAMPLE **A.** Check it out:

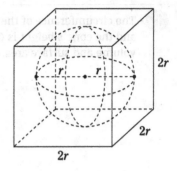

$$Vol_{Sphere} = \frac{4}{3}\pi r^3$$

$$Vol_{Cube} = (2r)^3$$

$$= 8r^3$$

$$\frac{Vol_{Sphere}}{Vol_{Cube}} = \frac{\frac{4}{3}\pi r^3}{8r^3}$$

$$= \frac{\pi}{6}$$

$$\approx 0.52$$

Thus, if you buy, say, a basketball that comes in a box, the basketball takes up about 52 percent of the volume of a box. (I'm sure you've been dying to know this.)

$$SA_{Sphere} = 4\pi r^2$$

$$SA_{Cube} = 6 \text{ sides} \cdot (2r)^2$$

$$= 24r^2$$

$$\frac{SA_{Sphere}}{SA_{Cube}} = \frac{4\pi r^2}{24r^2}$$

$$= \frac{\pi}{6} \text{ or about } 52\%$$

The very same percentage.

---

11 A cylinder with a radius of $\sqrt{5}$ and a height of 4 is inscribed in a sphere. Find the volume and surface area of the sphere.

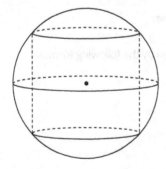

12 The hemispherical (half-sphere) top of a 50-foot-tall grain silo has a surface area of 200π square feet. How many cubic feet of grain can the silo hold?

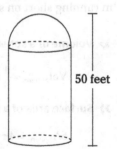

50 feet

Solutions

1. For both the volume and surface area, you need the sides of the $30°-60°-90°$ triangle. Its short leg is 6, so its long leg is $6\sqrt{3}$ and its hypotenuse is 12. You're all set to go.

$$\text{Vol} = \text{area}_{\text{base}} \cdot \text{height}$$
$$= \frac{1}{2}(6)(6\sqrt{3}) \cdot 10$$
$$= 18\sqrt{3} \cdot 10$$
$$= 180\sqrt{3} \text{ units}^3$$

$$\text{SA} = 2(\text{area}_{\text{base}}) + \text{three lateral rectangles}$$
$$= 2(18\sqrt{3}) + 6 \cdot 10 + 6\sqrt{3} \cdot 10 + 12 \cdot 10$$
$$= 36\sqrt{3} + 60 + 60\sqrt{3} + 120$$
$$= 180 + 96\sqrt{3} \text{ units}^2$$

2. All you need is the length of the base of the equilateral triangle. The triangle's altitude is 6, and that's the long leg of a $30°-60°-90°$ triangle. The short leg is therefore $\frac{6}{\sqrt{3}}$, or $2\sqrt{3}$, and the hypotenuse is twice that, or $4\sqrt{3}$ (see Chapter 4 for more on making this calculation). And that's the length, of course, of the sides of the equilateral triangle. Thus,

$$\text{Vol} = \text{area}_{\text{base}} \cdot \text{height}$$
$$= \frac{1}{2}(4\sqrt{3})(6) \cdot 12$$
$$= 144\sqrt{3} \text{ units}^3$$

$$\text{SA} = 2 \cdot \text{area}_{\text{base}} + \text{three lateral rectangles}$$
$$= 2(12\sqrt{3}) + 3(12 \cdot 4\sqrt{3})$$
$$= 24\sqrt{3} + 144\sqrt{3}$$
$$= 168\sqrt{3} \text{ units}^2$$

*3. Draw $\overline{CH}$; that's the hypotenuse of yet another $30°-60°-90°$ triangle ($\triangle CGH$). The length of the short leg, $\overline{CG}$, is 2, so CH is 4.

Now, note that $\triangle BCH$ is a right triangle with its right angle at C. One of its legs is 4 and its hypotenuse is 8, so — hold onto your hat — $\triangle BCH$ is *another* $30°-60°-90°$ triangle. The length of its long leg, $\overline{BC}$, is the length of the short leg, $\overline{CH}$, times $\sqrt{3}$, so BC is $4\sqrt{3}$. You have what you need to finish:

$$\text{Vol} = l \cdot w \cdot h \quad \text{(the same thing as } \text{area}_{\text{base}} \cdot \text{height)}$$
$$= 4\sqrt{3} \cdot 2\sqrt{3} \cdot 2$$
$$= 48 \text{ units}^3$$

$$\text{SA} = (2 \cdot \text{base}) + (2 \cdot \text{front}) + (2 \cdot \text{right side}) \quad \text{(the same thing as } 2 \cdot \text{area}_{\text{base}} + \text{lateral rectangles)}$$
$$= 2(4\sqrt{3} \cdot 2\sqrt{3}) + 2(4\sqrt{3} \cdot 2) + 2(2\sqrt{3} \cdot 2)$$
$$= 2 \cdot 24 + 2 \cdot 8\sqrt{3} + 2 \cdot 4\sqrt{3}$$
$$= 48 + 24\sqrt{3} \text{ units}^2$$

(4) Here are the answers:

a. The diameter is 4, so the radius is 2; thus,

$$\text{Vol} = \text{area}_{\text{base}} \cdot \text{height}$$
$$= \left(\pi r^2\right)(h)$$
$$= \left(\pi \cdot 2^2\right)(4)$$
$$= 16\pi \text{ units}^3$$

$$\text{SA} = 2 \cdot \text{area}_{\text{base}} + \text{lateral area} \quad (\text{which equals } \textit{circumference} \cdot \textit{height})$$
$$= 2\left(\pi r^2\right) + (2\pi r)(h)$$
$$= 2\left(\pi \cdot 2^2\right) + (2\pi \cdot 2)(4)$$
$$= 8\pi + 16\pi$$
$$= 24\pi \text{ units}^2$$

*b. This question is a great think-outside-the-box problem. Here's the trick: Imagine dividing the cylinder in half by cutting it along a plane that goes through A, B, and C and cuts the base along the dotted diameter. Now take the front half of the lateral area, uncurl it, and lay it flat. Here's what you get:

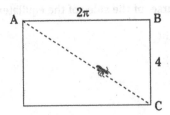

$\overline{AB}$ (in this rectangle, not across the top of the cylinder) has a length of half the circumference of the cylinder. That's 2π. The shortest path from A to C is straight, of course — that's $\overline{AC}$, which is the hypotenuse of right $\triangle ABC$. Its length is

$$c^2 = a^2 + b^2$$
$$= 4^2 + (2\pi)^2$$
$$= 16 + 4\pi^2$$
$$c = \sqrt{16 + 4\pi^2} = 2\sqrt{4 + \pi^2}$$
$$\approx 7.4$$

That's the shortest route; it curves along the outside of the cylinder, going diagonally down from A to C.

(5) You need the cylinder's radius and height. The diameter is 6, so the radius is 3. To get the height, you use the fact that the lateral area is a rectangle with an area of *circumference* · *height*. So

$$60\pi = \pi \cdot \text{diameter} \cdot \text{height}$$
$$60\pi = 6\pi h$$
$$10 = h$$

Thus,

$$\begin{aligned} \text{Vol} &= \text{area}_{\text{base}} \cdot \text{height} \\ &= \pi r^2 h \\ &= \pi \cdot 3^3 \cdot 10 \\ &= 90\pi \ \text{units}^3 \end{aligned}$$

And

$$\begin{aligned} \text{SA} &= 2 \cdot \text{area}_{\text{base}} + \text{lateral area} \\ &= 2\pi r^2 + 60\pi \quad \text{(this was given)} \\ &= 2\pi \cdot 3^2 + 60\pi \\ &= 78\pi \ \text{units}^2 \end{aligned}$$

(*6) You're given the surface area, so you have to begin with that formula:

$$\begin{aligned} \text{SA} &= 2 \cdot \text{area}_{\text{base}} + \text{lateral rectangle} \\ \text{SA} &= 2\pi r^2 + 2\pi rh \end{aligned}$$

Now plug in the given information:

$$\begin{aligned} 54\pi &= 2\pi r^2 + 2\pi r \cdot 6 \\ 54\pi &= 2\pi \left(r^2 + 6r \right) \\ 27 &= r^2 + 6r \\ r^2 + 6r - 27 &= 0 \\ (r+9)(r-3) &= 0 \\ r &= -9 \ \text{or} \ 3 \end{aligned}$$

You can reject −9, so r is 3. The rest is a walk in the park:

$$\begin{aligned} \text{Vol} &= \text{area}_{\text{base}} \cdot \text{height} \\ &= \pi r^2 h \\ &= \pi \cdot 3^2 \cdot 6 \\ &= 54\pi \ \text{units}^3 \end{aligned}$$

This cylinder is unusual and interesting because both the surface area and the volume are 54π. (For extra credit: Do you see why I didn't say that the surface area and the volume are *equal*?)

(7) To get the surface area, you need the slant heights, the lengths of $\overline{ZS}$ and $\overline{ZT}$. (Note that because this is not a regular pyramid — a pyramid with a regular polygon as its base and congruent lateral edges — these slant heights are not equal.) Then you use one of the slant heights to get the pyramid's height.

Keep looking for right triangles — that's the key to problems like this. $\triangle ASZ$ is a right triangle with a leg of 5 (half of $\overline{AB}$) and hypotenuse of $5\sqrt{10}$, so

$$\begin{aligned} (ZS)^2 + (AS)^2 &= (ZA)^2 \\ (ZS)^2 + 5^2 &= \left(5\sqrt{10} \right)^2 \\ (ZS)^2 &= 225 \\ ZS &= 15 \end{aligned}$$

$\triangle BTZ$ is another right triangle, with a leg of 9 (half the length of $\overline{BC}$):

$$(ZT)^2 + 9^2 = \left(5\sqrt{10}\right)^2$$
$$(ZT)^2 = 169$$
$$ZT = 13$$

You're all set to do the surface area:

$$SA_{Pointy\text{-}Top} = area_{base} + four\ lateral\ triangles$$

$$= 10 \cdot 18 + \underbrace{\frac{1}{2}(10)(15)}_{left\,face} + \underbrace{\frac{1}{2}(18)(13)}_{front} + \underbrace{\frac{1}{2}(10)(15)}_{right\,(same\ as\ left)} + \underbrace{\frac{1}{2}(18)(13)}_{back\,(same\ as\ front)}$$

$$= 180 + 75 + 117 + 75 + 117$$

$$= 564\ units^2$$

For the volume, you need the height, ZN. Well, $\triangle ZNS$ is a right triangle with a leg, $\overline{SN}$, that measures half of BC (so SN is 9) and a hypotenuse, $\overline{ZS}$, that's 15 units long. You can finish with the Pythagorean Theorem, or if you're on your toes, you'll notice that this triangle is in the $3:4:5$ family and that ZN is thus 12.

$$Volume_{Pointy\text{-}Top} = \frac{1}{3}\,area_{base} \cdot height$$

$$= \frac{1}{3}(10 \cdot 18) \cdot 12$$

$$= 720\ units^3$$

(*8) You can get the slant height, AE, fairly easily because $\triangle AEC$ is a $30° - 60° - 90°$ triangle. The short leg, $\overline{CE}$, is 3 units long, so AE is $3\sqrt{3}$.

To get the height, AF, imagine looking down on the base, $\triangle BCD$, like this:

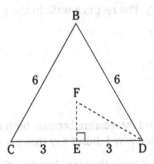

$\triangle EFD$ is, naturally, another $30° - 60° - 90°$ triangle. (You were expecting, maybe, a $29° - 57° - 94°$ triangle?) $\overline{ED}$ is half of edge $\overline{CD}$, so ED is 3, and that makes FE $\frac{3}{\sqrt{3}}$, or $\sqrt{3}$, and FD $2\sqrt{3}$.

Now you can use either $\overline{FE}$ with $\overline{AE}$ or $\overline{FD}$ with $\overline{AD}$ to get the height. I use $\overline{FE}$. Look back at the 3-D figure. $\triangle AFE$ is a right triangle, so you get AF with the Pythagorean Theorem:

$$(AF)^2 + (FE)^2 = (AE)^2$$
$$(AF)^2 + \left(\sqrt{3}\right)^2 = \left(3\sqrt{3}\right)^2$$
$$(AF)^2 + 3 = 27$$
$$AF = \sqrt{24} = 2\sqrt{6}$$

Time for the formulas, schmormulas:

$$\text{Vol}_{\text{Pointy-Top}} = \frac{1}{3} \, \text{area}_{\text{base}} \cdot \text{height}$$

$$= \frac{1}{3} \cdot \frac{s^2 \sqrt{3}}{4} \cdot h$$

$$= \frac{1}{3} \cdot \frac{6^2 \sqrt{3}}{4} \cdot 2\sqrt{6}$$

$$= 6\sqrt{18}$$

$$= 18\sqrt{2} \ \text{units}^3$$

For surface area, I just realized that you don't need to use the formula. Instead, you can just use the fact that a regular tetrahedron is four equilateral triangles. Thus,

$$\text{SA} = 4 \cdot \frac{6^2 \sqrt{3}}{4}$$

$$= 36\sqrt{3} \ \text{units}^2$$

9 **Circumference equals $2\pi r$, so**

$$16\pi = 2\pi r$$

$$r = 8$$

Now, the height of 6 and radius of 8 form the legs of a right triangle with the slant height as its hypotenuse. You have a triangle in the $3:4:5$ family, so the slant height is 10. Thus,

$$\text{Vol}_{\text{Pointy-Top}} = \frac{1}{3} \, \text{area}_{\text{base}} \cdot \text{height}$$

$$= \frac{1}{3} \, \pi r^2 h$$

$$= \frac{1}{3} \, \pi r \cdot 8^2 \cdot 6$$

$$= 128\pi \ \text{units}^3$$

$$\text{SA}_{\text{Pointy-Top}} = \text{area}_{\text{base}} + \text{one lateral "triangle"}$$

$$\left(\frac{1}{2} \, \text{circumference} \cdot \text{slant height} \right)$$

$$= \pi r^2 + \frac{1}{2} (2\pi r)(\text{slant height})$$

$$= \pi \cdot 8^2 + \frac{1}{2} (2\pi \cdot 8)(10)$$

$$= 64\pi + 80\pi$$

$$= 144\pi \ \text{units}^2$$

(10) Here are the answers:

*a. Call the height of the left-side cone x; then the height of the right-side cone is $21-x$.

$$\text{Total volume} = \text{vol}_{\text{left-side cone}} + \text{vol}_{\text{right-side cone}}$$

$$= \frac{1}{3} \cdot \text{area}_{\text{base}} \cdot \text{height} + \frac{1}{3} \cdot \text{area}_{\text{base}} \cdot \text{height}$$

$$= \frac{1}{3}\left(\pi \cdot 8^2\right)(x) + \frac{1}{3}\left(\pi \cdot 8^2\right)(21-x)$$

$$= \left(\frac{64\pi}{3}\right)(x) + \left(\frac{64\pi}{3}\right)(21-x)$$

$$= \frac{64\pi}{3}[x + (21-x)]$$

$$= \frac{64\pi}{3} \cdot 21$$

$$= 448\pi \text{ units}^3$$

The way the x drops out tells you that the x is irrelevant and, therefore, that the volume of this shape will be the same regardless of how far to the left or right the circular "base" is. Pretty nifty, eh?

b. The surface area, on the other hand, does depend on where the "base" is. The lateral area of a cone equals $\frac{1}{2}(circumference)(slant\ height)$, so for the left-side cone

$$80\pi = \frac{1}{2}(2\pi \cdot 8)(\text{slant height})$$

$$80\pi = 8\pi(\text{slant height})$$

$$\text{slant height} = 10$$

Will wonders never cease! You have another $3:4:5$ triangle here. So, the height of the left-side cone is 6. Then, $21-6$ gives you 15, the height of the right-side cone. And then you notice, of course, that you have an $8-15-17$ triangle on the right, so the right-side slant height is 17.

$$SA = \frac{1}{2}(\text{circumference})(\text{slant height})$$

$$= \frac{1}{2}(2\pi \cdot 8)(17)$$

$$= 136\pi \text{ units}^2$$

(11) In many sphere problems (like with many circle problems), the key is finding the right radius or radii. Often, a radius becomes the hypotenuse of a right triangle. Find the right one? Here it is:

$$r^2 = 2^2 + \sqrt{5}^2$$
$$r^2 = 9$$
$$r = 3$$

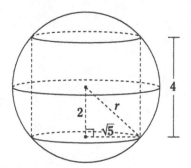

Thus,

$$\text{Vol}_{\text{Sphere}} = \frac{4}{3}\pi r^3$$
$$= \frac{4}{3}\pi \cdot 3^3$$
$$= 36\pi \text{ units}^3$$

$$\text{SA}_{\text{Sphere}} = 4\pi r^2$$
$$= 4\pi \cdot 3^2$$
$$= 36\pi \text{ units}^2$$

Only a sphere with a radius of 3 has a volume (in cubic units) equal to its surface area (in square units).

(12) The surface area of a sphere equals $4\pi r^2$, so, obviously, a hemisphere has a surface area of half that, or $2\pi r^2$:

$$\text{SA}_{\text{Hemisphere}} = 2\pi r^2$$
$$200\pi = 2\pi r^2$$
$$100 = r^2$$
$$r = 10$$

The radius of the cylinder is also 10, of course.

The "height" of the hemisphere (from its "peak" straight down to the center of its circular base) is just one of its radii, so that's 10. Because the total height is 50, the height of the cylinder is 50 – 10, or 40. Now you have what you need to finish:

$$\text{Total volume} = \text{vol}_{\text{cylinder}} + \text{vol}_{\text{hemisphere}}$$
$$= \text{area}_{\text{base}} \cdot \text{height} + \frac{1}{2}\text{vol}_{\text{sphere}}$$
$$= \pi \cdot 10^2 \cdot 40 + \frac{1}{2}\left(\frac{4}{3}\pi \cdot 10^3\right)$$
$$= 4000\pi + \frac{2000\pi}{3}$$
$$= \frac{14,000\pi}{3}$$
$$\approx 14,661 \text{ feet}^3$$

6

Coordinate Geometry, Loci, and Constructions: Proof and Non-Proof Problems

Chapter **13**

Coordinate Geometry, Courtesy of Descartes (Including Proofs)

F or someone who is said to have slept till 11 a.m. every day, René Descartes (1596–1650) — *not* pronounced "Dess-cart-eez" — sure achieved a lot: world-famous philosopher, music theorist, physicist, and, of course, mathematician. Not too shabby, eh? Of interest here is the fact that he played a significant role in the evolution of geometry: He made the move from analyzing geometric shapes that exist independently of any location or orientation (the way the Greeks did geometry, and the way I've done problems up to this point in this book) to placing geometric shapes in the *x-y* coordinate system and using algebra to analyze them.

Formulas, Schmormulas: Slope, Distance, and Midpoint

REMEMBER

Here are a few formulas you probably know (have a faint recollection of?) from Algebra I. You'll use these formulas to do the same sort of problems you have done in earlier chapters but in a completely different way.

>> **Slope formula:** The slope of a line containing two points — (x_1, y_1) and (x_2, y_2) — is given by the following formula (don't ask me why, but the letter m is typically used for the slope):

$$\text{Slope} = m = \frac{y_2 - y_1}{x_2 - x_1} = \frac{\text{rise}}{\text{run}}$$

>> **Slope of horizontal lines:** The slope of a *horizontal* line is zero. Think about driving on a horizontal, flat road — the road has no steepness or slope.

>> **Slope of vertical lines:** The slope of a *vertical* line is *undefined* (because the run is zero and you can't divide by zero). Think about driving up a vertical road — you can't do it; it's impossible. And it's impossible to compute the slope of a vertical line.

>> **Slope of parallel lines:** The slopes of parallel lines are equal (unless both lines are vertical, in which case both of their slopes are undefined).

>> **Slope of perpendicular lines:** The slopes of perpendicular lines are opposite reciprocals of each other, like 3 and $-\frac{1}{3}$ or $\frac{2}{5}$ and $-\frac{5}{2}$ (unless one line is horizontal [slope = 0] and the other line is vertical [slope is undefined]).

>> **Midpoint formula:** The midpoint of the segment with endpoints at (x_1, y_1) and (x_2, y_2) is given by the formula

$$\text{Midpoint} = \left(\frac{x_1 + x_2}{2}, \frac{y_1 + y_2}{2} \right)$$

Just remember, the midpoint is the average of the *x*'s and the average of the *y*'s.

>> **Distance formula:** The distance from (x_1, y_1) to (x_2, y_2) is given by the following formula:

$$\text{Distance} = \sqrt{(x_2 - x_1)^2 + (y_2 - y_1)^2}$$

The distance formula is simply the Pythagorean Theorem solved for *c*, the hypotenuse. The legs of the right triangle have lengths equal to the change in the *x*-coordinates and the change in the *y*-coordinates. If you just remember this connection, you can always solve a distance problem with the Pythagorean Theorem even if you forget the distance formula.

EXAMPLE

Q. Show that *ISOT* is an isosceles trapezoid.

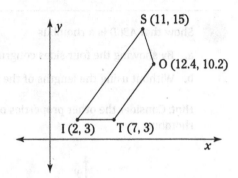

A. To prove that *ISOT* is an isosceles trapezoid, you must show that $\overline{IS} \parallel \overline{TO}$ (definition of a trapezoid; see Chapter 5) and that $\overline{IT} \cong \overline{SO}$ (the meaning of isosceles). (And for sticklers, there's one more thing to show: It's totally obvious from the diagram, but you have to show that $\overline{IT}$ is *not* parallel to $\overline{SO}$ — otherwise, *ISOT* would be a parallelogram and thus not a trapezoid.)

First, check the slopes:

$$\text{Slope}_{\overline{IS}} = \frac{15-3}{11-2} = \frac{12}{9} = \frac{4}{3}$$

$$\text{Slope}_{\overline{TO}} = \frac{10.2-3}{12.4-7} = \frac{7.2}{5.4} = \frac{4}{3}$$

Check; $\overline{IS}$ is parallel to $\overline{TO}$.

Now check the lengths of $\overline{IT}$ and $\overline{SO}$ with the distance formula. Actually, although the distance formula works fine for *IT*, you don't need it. For vertical and horizontal segments, the distance is obvious. From *I* to *T*, you go straight across from 2 to 7, so the length is 5. For *SO*, you have

$$
\begin{aligned}
SO &= \sqrt{(12.4-11)^2 + (10.2-15)^2} \\
&= \sqrt{(1.4)^2 + (-4.8)^2} \\
&= \sqrt{1.96 + 23.04} \\
&= \sqrt{25} \\
&= 5
\end{aligned}
$$

Check; $\overline{IT} \cong \overline{SO}$.

You can easily check for yourself that $\overline{IT}$ is not parallel to $\overline{SO}$, so that does it: *ISOT* is an isosceles trapezoid.

 **Show that** *ABCD* **is a rhombus**

    a. By showing the four sides congruent

    b. Without using the lengths of the sides

    Hint: Consider the other properties of a rhombus.

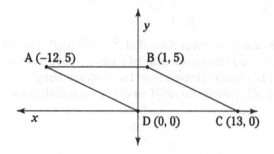

 Using the diagram,

    a. Show that *PLOG* is a parallelogram

    b. Find its area and perimeter

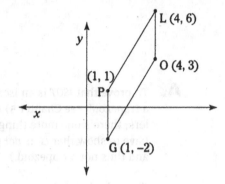

 Take a look at △*ABC*.

    a. What type of triangle is △*ABC*: acute, obtuse, or right? Equilateral, isosceles, or scalene?

    b. Find its area and perimeter

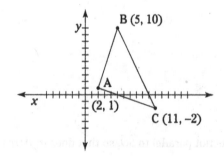

 Use the diagram and its labeled coordinates to

    a. Show that *KITE* is a kite

    b. Find its area

    c. Find the point where its diagonals intersect

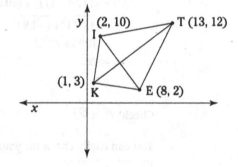

Mastering Coordinate Proofs with Algebra

In this section, you get to see the power of doing geometry *analytically,* that is, with algebra. You prove the same types of things you proved in the chapter on quadrilaterals, but this time without any of the methods you used there (like congruent triangles, CPCTC, alternate interior angles, and so on). Sometimes proving something analytically is easier than with two-column proof methods. The second practice problem (#6) is a case in point. If you happen to see the trick, doing the proof the regular two-column way isn't that hard. But if you don't, you may not be able to do the proof the regular way. Doing it analytically, however, works like a charm.

Q. Use the isosceles trapezoid in the figure to prove that the diagonals in an isosceles trapezoid are congruent.

EXAMPLE

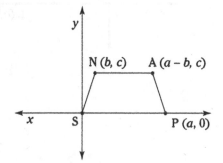

Note: I can't explain it fully here, so you have to take my word for it that this figure covers all conceivable isosceles trapezoids. You can place one vertex at the origin and another on the x-axis at $(a, 0)$ and put the whole trapezoid in the first quadrant "with no loss of generality," as mathematicians say. (*Caution:* It's probably not the best idea to use this phrase when you're out on a date.)

A. The proof is sort of one step long (or one idea long): You simply use the distance formula to show that the diagonals are congruent (for this property of isosceles trapezoids and more, check out Chapter 7):

$$SA \overset{?}{=} NP$$

$$\sqrt{(a-b-0)^2 + (c-0)^2} \overset{?}{=} \sqrt{(a-b)^2 + (0-c)^2}$$

$$\sqrt{(a-b)^2 + c^2} \overset{?}{=} \sqrt{(a-b)^2 + (-c)^2}$$

You know that c^2 is the same as $(-c)^2$, so these values are equal. That does it.

5 Given that the quadrilateral in the figure is a parallelogram, prove analytically that the diagonals of a parallelogram bisect each other.

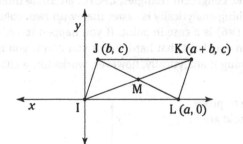

6 Use the figure to prove that if you connect the midpoints of the sides of any quadrilateral, you create a parallelogram. (For an extra challenge, try to prove this with ordinary two-column proof methods.)

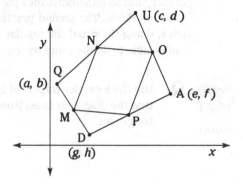

Using the Equations of Lines and Circles

Lines and circles don't seem to have much in common at first glance; after all, lines are one-dimensional objects that go one forever in both directions, while circles are two-dimensional objects (if you count their interiors) that cover a definite amount of space. The major thing that lines and circles *do* have in common is that they become very important in subsequent math classes, like trigonometry and calculus. These equations will keep popping up in your classes over and over again, so you might as well get used to 'em now and get ahead of the game.

REMEMBER

Line equations. Here are the basic forms for equations of lines.

> » **Slope-intercept form:**
>
> $y = mx + b$
>
> where m is the slope and b is the y-intercept $(0, b)$.

> » **Point-slope form:**
>
> $y - y_1 = m(x - x_1)$
>
> where m is the slope and (x_1, y_1) is a point on the line.

>> **Horizontal line:**

$$y = b$$

where b is the y-intercept.

>> **Vertical line:**

$$x = a$$

where a is the x-intercept.

REMEMBER

Circle equation. And here's the equation of a circle:

$$(x - h)^2 + (y - k)^2 = r^2$$

where (h, k) is the center of the circle and r is its radius.

EXAMPLE

Q. A circle whose center is at $(6, 5)$ is tangent to a line at $(2, 7)$. What are the equations of the circle and the line, and what is the line's y-intercept?

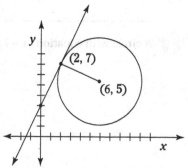

A. You have the circle's center, so all you need for the circle's equation is its radius. Use the distance formula:

$$r = \sqrt{(6 - 2)^2 + (5 - 7)^2}$$
$$= \sqrt{4^2 + (-2)^2}$$
$$= \sqrt{20}$$
$$= 2\sqrt{5}$$

Thus, the equation of the circle is

$$(x - 6)^2 + (y - 5)^2 = \left(2\sqrt{5}\right)^2, \text{ or}$$

$$(x - 6)^2 + (y - 5)^2 = 20$$

For the equation of the line, you have a point, so all you need is the slope. A line tangent to a circle is perpendicular to the radius drawn to the point of tangency, so first you need the slope of this particular radius:

$$\text{Slope}_{\text{Radius}} = \frac{7 - 5}{2 - 6} = -\frac{1}{2}$$

Because the line is perpendicular to the radius, their slopes are opposite reciprocals. The opposite reciprocal of $-\frac{1}{2}$ is 2, so that's the line's slope, and now you have everything you need to plug into the point–slope form:

$$y - y_1 = m(x - x_1)$$

$$y - 7 = 2(x - 2)$$

Finally, to get the y-intercept, just transform this equation into slope–intercept form:

$$y - 7 = 2(x - 2)$$
$$y - 7 = 2x - 4$$
$$y = 2x + 3$$

The y-intercept is (0, 3).

---

*7 A circle with equation $(x - 7)^2 + y^2 = r^2$ is tangent to lines at (4, 4) and (11, −3). Find r and (a, b).

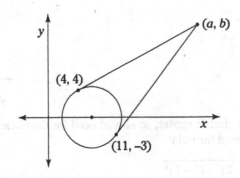

Solutions

1. Here's what you do:

a. From A to B, you go straight across from -12 to 1, so AB is 13. And DC is obviously also 13. Now use the distance formula for AD and BC (though you don't need to if you recognize the $5-12-13$ triangles — see Chapter 4):

$$AD = \sqrt{[0-(-12)]^2 + (0-5)^2}$$
$$= \sqrt{12^2 + (-5)^2}$$
$$= 13$$

$$BC = \sqrt{(13-1)^2 + (0-5)^2}$$
$$= \sqrt{12^2 + (-5)^2}$$
$$= 13$$

That's it. All four sides have a length of 13, so $ABCD$ is a rhombus.

b. You can show that $ABCD$ is a rhombus without using the lengths of the sides by first showing that $ABCD$ is a parallelogram and then that its diagonals are perpendicular (check out Chapter 7 for the properties of a rhombus).

$\overline{AB}$ and $\overline{DC}$ have slopes of 0, so they're parallel. Now check the slopes of $\overline{AD}$ and $\overline{BC}$:

$$\text{Slope}_{\overline{AD}} = \frac{0-5}{0-(-12)} = -\frac{5}{12}$$

$$\text{Slope}_{\overline{BC}} = \frac{0-5}{13-1} = -\frac{5}{12}$$

With two pairs of parallel sides, $ABCD$ must be a parallelogram.

Now check the slopes of the diagonals:

$$\text{Slope}_{\overline{AC}} = \frac{0-5}{13-(-12)} = -\frac{1}{5}$$

$$\text{Slope}_{\overline{DB}} = \frac{5-0}{1-0} = 5$$

Because 5 and $-\frac{1}{5}$ are opposite reciprocals, $\overline{AC} \perp \overline{DB}$. $ABCD$ is thus a rhombus, because a parallelogram with perpendicular diagonals is a rhombus.

2. Here's how this one unfolds:

a. $\overline{PG}$ and $\overline{LO}$ are both vertical, so they're parallel. Now check the other sides:

$$\text{Slope}_{\overline{PL}} = \frac{6-1}{4-1} = \frac{5}{3}$$

$$\text{Slope}_{\overline{GO}} = \frac{3-(-2)}{4-1} = \frac{5}{3}$$

That's all there is to it. $PLOG$ is a parallelogram.

b. You can use $\overline{PG}$ for the base of the parallelogram; its length is 3. The height of *PLOG* is thus horizontal because it's perpendicular to the base, $\overline{PG}$; it goes straight to the right from $x = 1$ to $x = 4$. So, the height is also 3, and the area of *PLOG* is thus 3 times 3 (base times height), or 9 units^2 (see Chapter 7 for more on calculating the area of quadrilaterals).

The perimeter is a snap. $\overline{PG}$ and $\overline{LO}$ both have a length of 3. And

$$PL = \sqrt{(4-1)^2 + (6-1)^2}$$
$$= \sqrt{3^2 + 5^2}$$
$$= \sqrt{34}$$

Because you already know that *PLOG* is a parallelogram, $\overline{GO}$ has to be congruent to $\overline{PL}$, so it's also $\sqrt{34}$ units long. Thus, the perimeter of *PLOG* is $3 + 3 + \sqrt{34} + \sqrt{34}$, or $6 + 2\sqrt{34}$.

(3) To solve these problems, you use the triangle basics I cover in Chapter 4.

a. $\angle A$ looks like a right angle, so cross your fingers and check the slopes of $\overline{AB}$ and $\overline{AC}$ (if $\angle A$ is a right angle, this problem becomes much easier).

$$\text{Slope}_{\overline{AB}} = \frac{10-1}{5-2} = 3$$

$$\text{Slope}_{\overline{AC}} = \frac{-2-1}{11-2} = -\frac{1}{3}$$

These answers are opposite reciprocals, so $\overline{AB} \perp \overline{AC}$, and thus $\angle A$ is a right angle; $\triangle ABC$ is a right triangle.

Now compute the lengths of legs $\overline{AB}$ and $\overline{AC}$:

$$AB = \sqrt{(5-2)^2 + (10-1)^2}$$
$$= \sqrt{3^2 + 9^2}$$
$$= \sqrt{90}$$
$$= 3\sqrt{10}$$

$$AC = \sqrt{(11-2)^2 + (-2-1)^2}$$
$$= \sqrt{9^2 + (-3)^2}$$
$$= 3\sqrt{10}$$

$\overline{AB} \cong \overline{AC}$, so voilà, you have a $45° - 45° - 90°$ triangle, or, in other words, an isosceles right triangle.

b. The area of a right triangle equals one half the product of its legs, so

$$\text{Area}_{\triangle ABC} = \frac{1}{2}\left(3\sqrt{10}\right)\left(3\sqrt{10}\right)$$
$$= 45 \text{ units}^2$$

The legs are $3\sqrt{10}$, so the hypotenuse is $\sqrt{2} \cdot 3\sqrt{10}$, or $6\sqrt{5}$, and thus the perimeter of $\triangle ABC$ is $3\sqrt{10} + 3\sqrt{10} + 6\sqrt{5}$, or $6\sqrt{10} + 6\sqrt{5}$.

④ If you need a review, Chapter 6 explains the properties of kites.

a. By definition, a kite must have two pairs of adjacent congruent sides. Use the distance formula:

$$KI = \sqrt{(2-1)^2 + (10-3)^2}$$
$$= \sqrt{1^2 + 7^2}$$
$$= 5\sqrt{2}$$

$$KE = \sqrt{(8-1)^2 + (2-3)^2}$$
$$= 5\sqrt{2}$$

So far, so good.

$$IT = \sqrt{(13-2)^2 + (12-10)^2}$$
$$= \sqrt{11^2 + 2^2}$$
$$= 5\sqrt{5}$$

$$ET = \sqrt{(13-8)^2 + (12-2)^2}$$
$$= \sqrt{5^2 + 10^2}$$
$$= 5\sqrt{5}$$

Bingo. *KITE* is a kite.

b. The area of a kite equals half the product of its diagonals (see Chapter 7), so you need their lengths:

$$KT = \sqrt{(13-1)^2 + (12-3)^2}$$
$$= \sqrt{12^2 + 9^2}$$
$$= 15$$

$$IE = \sqrt{(8-2)^2 + (2-10)^2}$$
$$= \sqrt{6^2 + (-8)^2}$$
$$= 10$$

$$\text{Area}_{KITE} = \frac{1}{2}d_1 d_2$$
$$= \frac{1}{2}(15)(10)$$
$$= 75 \text{ units}^2$$

c. $\overline{KT}$ is the perpendicular bisector of $\overline{IE}$ (property of a kite; see Chapter 6), so all you need is $\overline{IE}$'s midpoint:

$$\text{Midpoint}_{\overline{IE}} = \left(\frac{x_1 + x_2}{2}, \frac{y_1 + y_2}{2} \right)$$

$$= \left(\frac{2+8}{2}, \frac{10+2}{2} \right)$$

$$= (5, 6)$$

5. The proof here is odd in a way, but it works. You might think that you have to first find where the diagonals cross and then show that this point bisects each diagonal. Instead, you simply show that the midpoints of the two diagonals are at the same point:

$$\text{Midpoint}_{\overline{JL}} = \left(\frac{b+a}{2}, \frac{c+0}{2} \right)$$

$$= \left(\frac{b+a}{2}, \frac{c}{2} \right)$$

$$\text{Midpoint}_{\overline{IK}} = \left(\frac{0+a+b}{2}, \frac{0+c}{2} \right)$$

$$= \left(\frac{a+b}{2}, \frac{c}{2} \right)$$

You're done. This simple procedure does, in fact, prove that the diagonals of any parallelogram bisect each other.

6. You need to get the coordinates of the midpoints using — hold onto your hat — the midpoint formula. Then use them to find the slopes of the sides of *MNOP*.

$$M = \left(\frac{a+g}{2}, \frac{b+h}{2} \right)$$

$$N = \left(\frac{a+c}{2}, \frac{b+d}{2} \right)$$

$$O = \left(\frac{c+e}{2}, \frac{d+f}{2} \right)$$

$$P = \left(\frac{g+e}{2}, \frac{h+f}{2} \right)$$

$$\text{Slope}_{\overline{MN}} = \frac{\frac{b+d}{2} - \frac{b+h}{2}}{\frac{a+c}{2} - \frac{a+g}{2}}$$

$$= \frac{(b+d)-(b+h)}{(a+c)-(a+g)} \quad \text{(multiplying top and bottom by 2)}$$

$$= \frac{d-h}{c-g}$$

$$\text{Slope}_{\overline{PO}} = \frac{\frac{d+f}{2} - \frac{h+f}{2}}{\frac{c+e}{2} - \frac{g+e}{2}}$$

$$= \frac{d-h}{c-g}$$

One pair of parallel sides down, one to go:

$$\text{Slope}_{\overline{MP}} = \frac{\frac{h+f}{2} - \frac{b+h}{2}}{\frac{g+e}{2} - \frac{a+g}{2}}$$

$$= \frac{f-b}{e-a}$$

$$\text{Slope}_{\overline{NO}} = \frac{\frac{d+f}{2} - \frac{b+d}{2}}{\frac{c+e}{2} - \frac{a+c}{2}}$$

$$= \frac{f-b}{e-a}$$

Bingo. It's a parallelogram. Pretty cool, eh? No matter what weird quadrilateral you begin with, you always get a parallelogram.

(*7) The circle's equation gives you its center: (7, 0). Now use the distance formula to get the radius:

$$r = \sqrt{(4-7)^2 + (4-0)^2}$$

$$= \sqrt{(-3)^2 + 4^2}$$

$$= \sqrt{25} = 5$$

Well, bust my britches and bless my soul — another 3-4-5 triangle! What are the odds of that? To find (a, b), you need the equations of the tangent lines, and for that you need the slopes of the lines:

$$\text{Slope}_{\text{Radius to (4, 4)}} = \frac{4-0}{4-7} = -\frac{4}{3}$$

The tangent line is perpendicular to this radius, so its slope is the opposite reciprocal of $-\frac{4}{3}$, namely $\frac{3}{4}$. And now you have what you need for the point-slope form:

$$y - 4 = \frac{3}{4}(x - 4)$$

Use the same process for the other tangent line:

$$\text{Slope}_{\text{Radius to (11, -3)}} = \frac{-3-0}{11-7} = -\frac{3}{4}$$

The tangent line's slope is the opposite reciprocal of that, namely $\frac{4}{3}$, and thus its equation is

$$y - (-3) = \frac{4}{3}(x - 11)$$

Now find the point of intersection of the two lines by solving the system of equations with two unknowns. First solve each equation for y:

$$y - 4 = \frac{3}{4}(x - 4) \qquad y - (-3) = \frac{4}{3}(x - 11)$$

$$y = \frac{3}{4}(x - 4) + 4 \qquad y = \frac{4}{3}(x - 11) - 3$$

Now set the equations equal to each other and solve:

$$\frac{3}{4}(x - 4) + 4 = \frac{4}{3}(x - 11) - 3$$
$$9(x - 4) + 48 = 16(x - 11) - 36$$
$$9x - 36 + 48 = 16x - 176 - 36$$
$$9x + 12 = 16x - 212$$
$$224 = 7x$$
$$x = 32$$

Plugging this answer into either tangent line gives you a y-value of 25. Thus (a, b) is $(32, 25)$.

IN THIS CHAPTER

» A few reflections on reflections

» Shifting shapes with translations

» You spin me right round, Polly: Rotating polygons

» Reflecting thrice: Glide reflections

Chapter 14

Transforming the (Geometric) World: Reflections, Rotations, and Translations (No Proofs)

Y ou can take any figure, say a triangle, and use a *transformation* to move it or change it in some way. You can slide it, flip it over, shrink it or blow it up, warp it into a different shape, and so on. In this chapter, you practice problems involving transformations that don't change the size or shape of a figure. Such transformations — called *isometries* — take a figure and move it, or *map* it, onto a congruent figure. The "before" figure is called the *pre-image,* and the "after" figure is called the *image.*

Reflections on Mirror Images

I begin this isometries journey with reflections, not because they're the simplest subject you need to tackle here but because they're the building blocks of all other isometries. In fact, you can use a series of reflections to perform all the other transformations I discuss later in this chapter. For example, in the next section, I show you that you can translate a figure in any direction (which you could do by just sliding it) by instead reflecting the figure over one line and then reflecting it again over another line. In fact, if you take, say, two congruent triangles and place them anywhere in the *x-y* coordinate system — one flipped over, if you like, and rotated to any angle — and you want to map one of the triangles onto the other by a series of transformations, you never have to rotate or slide the triangle. In one, two, or three reflections (you never need more than three), you can make the "before" triangle land exactly on the "after" triangle. I find this result interesting and somewhat surprising.

A couple more things before working through an example. (Egad! A sentence fragment!) First, check out Figure 14-1. △*ABC* has been reflected over line *l*. The result is congruent △*PQR*. △*ABC* has also been slid to the right (that move is a *translation*, if you were wondering), producing congruent △*XYZ*. △*PQR* and △*XYZ* are congruent, but there's a basic difference between them: their orientation. Figures like △*ABC* and △*XYZ* have the same *orientation* because you can make one stack perfectly on top of the other by sliding and/or rotating it onto the other. Figures like △*ABC* and △*PQR*, on the other hand, have opposite orientations because you can't possibly get △*ABC* to line up with △*PQR* without flipping △*ABC* over. Read on for some theorems.

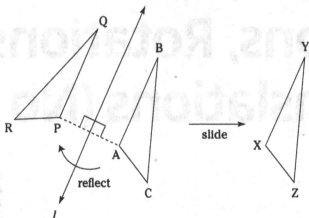

FIGURE 14-1:
A triangle
and its
transformations.

**THEOREMS &
POSTULATES**

Reflections and orientation:

» Reflecting a figure switches its orientation.

» If you reflect a figure and then reflect it again over the same line or a different line, the figure returns to its original orientation. More generally, if you reflect a figure an *even* number of times, the final result is a figure with the *same orientation*.

» Reflecting a figure an *odd* number of times produces a figure with the *opposite orientation*.

And here's one more thing about Figure 14-1. If you form $\overline{AP}$ by connecting pre-image A with its image point P (or B with Q or C with R), the reflecting line, l, is the perpendicular bisector of $\overline{AP}$. Pretty cool, huh?

Reflecting lines and connecting segments: When a figure is reflected, the *reflecting line* is the perpendicular bisector of all segments connecting points of the pre-image to corresponding points of the image.

THEOREMS & POSTULATES

After each transformation, you can label the image points with the *prime* symbol ('). If A is the pre-image, the image point is A'.

EXAMPLE

Q. A transformation T maps (or sends) all points (x, y) to (y, x). Symbolically, $T(x, y) = (y, x)$. This transformation is a reflection. Given the coordinates of the vertices of $\triangle ABC$, find the coordinates of the reflection of $\triangle ABC$, which is $\triangle A'B'C'$, and find the equation of the reflecting line.

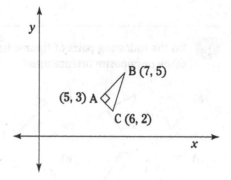

A. For vertex A, $T(5, 3) = (3, 5)$; that's A'

For B, $T(7, 5) = (5, 7)$; that's B'

For C, $T(6, 2) = (2, 6)$; that's C'

Now sketch $\triangle ABC$, $\triangle A'B'C'$, and the reflecting line.

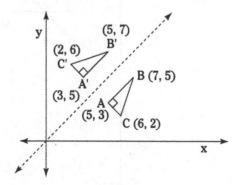

The reflecting line is the perpendicular bisector of $\overline{AA'}$ (and $\overline{BB'}$ and $\overline{CC'}$). To find this line, you first need the midpoint of $\overline{AA'}$; that's $\left(\dfrac{5+3}{2}, \dfrac{3+5}{2} \right)$ or $(4, 4)$. Next, compute the slope of $\overline{AA'}$ (see Chapter 13 for more on slope and midpoints); that's $\dfrac{5-3}{3-5} = -1$.

The perpendicular bisector is, of course, perpendicular to $\overline{AA'}$, so its slope is the opposite reciprocal of −1, which is 1. You have a point, (4, 4), and the slope, 1, of the perpendicular bisector, so you're all set to plug into the point-slope form (see Chapter 13 for more on line equations):

$$y - y_1 = m(x - x_1)$$
$$y - 4 = 1(x - 4)$$
$$y - 4 = x - 4$$
$$y = x$$

That's it.

---

1 Do the following pairs of figures have the same or opposite orientations?

a)

b)

c)

d)

e)

f)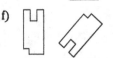

2 Reflect QRST over the line $y = x$.

a. Sketch $Q'R'S'T'$, and give the coordinates of Q' and R'.

b. What shape is $QQ'R'R$?

c. What's the area and perimeter of $QQ'R'R$?

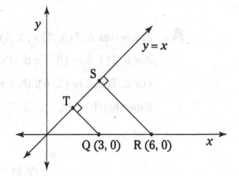

Sketch the reflected images and give the coordinates of the following triangles.

a. $\triangle ABC$ reflected over $y = x$ to $\triangle A'B'C'$

b. $\triangle A'B'C'$ reflected over $y = -x$ to $\triangle A''B''C''$

c. $\triangle A''B''C''$ reflected over the y-axis to $\triangle A'''B'''C'''$

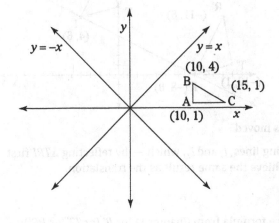

 Reflect $\triangle TUV$ over the line $y = 3x + 2$.

a. Find the coordinates of $\triangle T'U'V'$.

Hint: You need the equations of $\overline{TT'}$, $\overline{UU'}$ and $\overline{VV'}$.

b. Show that $\triangle T'U'V' \cong \triangle TUV$.

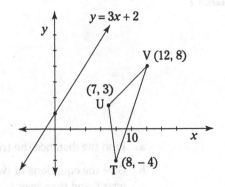

Lost in Translation

Translating or sliding a figure is probably the simplest transformation to picture. It's so simple, in fact, that there wouldn't be much to say about it if it weren't for the fact that you can produce a translation with two reflections. You can picture how this works by imagining that you have a playing card — say, the ace of spades — face up in front of you on a table. Now, grab the bottom edge of the card and flip the card over (going up, away from you), leaving the top edge of the card where it is. You should now see a face-down card whose bottom edge (the one close to you) is where the top edge was before you flipped it. Got it? If you repeat this flipping procedure, you should see the face-up ace again, pointing the same direction, and the card is now farther away from you by a distance equal to twice the height of the card. Thus, you see how two reflections (or flips) equals a slide.

A translation equals two reflections. A translation of a given distance along a given line is equivalent to two reflections over parallel lines that are perpendicular to the given line and separated by a distance equal to half the distance of the translation. As long as the parallel reflecting lines are separated by this distance, they can be located anywhere along the given line.

EXAMPLE

Q. The translation $(x, y) \rightarrow (x - 12, y - 6)$ maps $\triangle TRI$ to $\triangle T'R'I'$.

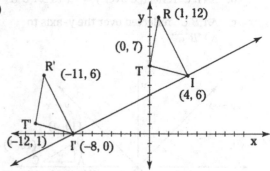

a. Find the distance the triangle has moved.

b. Give the equations of two reflecting lines, l_1 and l_2, which — by reflecting $\triangle TRI$ first over l_1 and then over l_2 — will achieve the same result as the translation.

A. Here's how it all goes down:

a. Piece o' cake. Just use the distance formula from Chapter 13 for II' (or TT' or RR'):

$$II' = \sqrt{(-8 - 4)^2 + (0 - 6)^2}$$
$$= \sqrt{144 + 36}$$
$$= 6\sqrt{5}$$

You can use a slight shortcut here if you realize that the translation instructions tell you that you're moving the figure 12 left and 6 down. If you see that, you just do $distance = \sqrt{12^2 + 6^2}$, and so on.

b. You need two parallel lines perpendicular to $\overline{II'}$ and separated by half the length of $\overline{II'}$. There are, literally, an infinite number of correct answers. Here's an easy way to find a pair of lines that work:

The pair of lines must be perpendicular to $\overline{II'}$, which has a slope of $\dfrac{6-0}{4-(-8)}$, or $\dfrac{1}{2}$, so the slope of the parallel lines is the opposite reciprocal of that, namely -2. The first line, l_1, can go through point I at $(4, 6)$. Its equation is thus

$$y - 6 = -2(x - 4)$$
$$y = -2x + 14$$

Make the second line, l_2, parallel to l_1 (so its slope is also -2) and have it go through the midpoint of $\overline{II'}$. With this choice, you make the distance between l_1 and l_2 the required distance — half the length of $\overline{II'}$. The midpoint of $\overline{II'}$ is $\left(\dfrac{4 + (-8)}{2}, \dfrac{6 + 0}{2} \right)$ or

(−2, 3). And thus, when you plug those numbers into the point-slope form of a line (see Chapter 13), the equation of l_2 is

$$y - 3 = -2[x - (-2)]$$
$$y = -2x - 1$$

Finito.

---

5 The translation $(x, y) \rightarrow (x, y + 5)$ maps *ISOC* onto *TRAP*. Find the equations of two reflecting lines that achieve the same result. Give three answers (in other words, three possible *pairs* of reflecting lines).

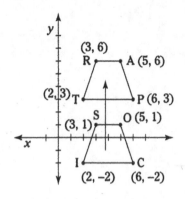

6 The translation $(x, y) \rightarrow (x + 9, y + 2)$ maps $\triangle ABC$ onto $\triangle A'B'C'$. Find a pair of parallel reflecting lines that achieves the same result.

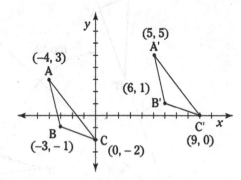

So You Say You Want a . . . Rotation?

I can't give you a revolution, but I have a bunch of rotations waiting for you in this section. You know what *rotation* means, of course, but one thing you may not realize about rotation transformations is that they include not only spinning a figure where it is, but also making it sort of move along an orbit centered at a point away from the figure (as in Figure 14-2). It might be more accurate to call this type of transformation a *revolution* instead of a rotation, but who am I to question the age-old terminology of geometry?

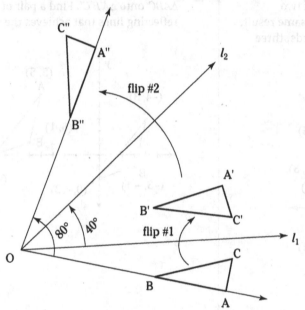

FIGURE 14-2:
A rotation is equivalent to two reflections.

A rotation, just like a translation, can be achieved by a pair of reflections. Look at Figure 14-2.

You can see that △*ABC* has been rotated 80° to △*A″B″C″*. Point *O* is called the *center of rotation*. It turns out that this same transformation can be achieved by reflecting △*ABC* over l_1 to △*A′B′C′* and then reflecting △*A′B′C′* over l_2 to △*A″B″C″*. The reflecting lines must pass through the center of rotation, and the angle between them must be half the angle of rotation. Pretty nifty, eh?

A rotation equals two reflections. A rotation through a given angle around a center of rotation is equivalent to two reflections over lines passing through the center of rotation and forming an angle half the measure of the angle of rotation.

THEOREMS & POSTULATES

Q. △*DEF* has been rotated counterclockwise onto △*D'E'F'*. Find the center of rotation.

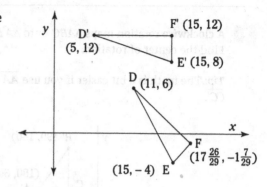

EXAMPLE

A. I haven't mentioned this process yet, so here I show you how to find a center of rotation. The trick is to use perpendicular bisectors. For this problem, the center of rotation lies at the intersection of the perpendicular bisectors of $\overline{DD'}$, $\overline{EE'}$, and $\overline{FF'}$. You need only two of these perpendicular bisectors, so use $\overline{DD'}$ and $\overline{EE'}$. (If you love working with fractions like $17\frac{26}{29}$, $\overline{FF'}$ would work as well.) The perpendicular bisector of $\overline{DD'}$ goes through its midpoint, which is $\left(\dfrac{11+5}{2}, \dfrac{6+12}{2}\right)$ or (8, 9). The slope of $\overline{DD'}$ is $\dfrac{12-6}{5-11}$, or –1, so the slope of the perpendicular bisector is the opposite reciprocal of that, which is 1. Write the equation of the line in point-slope form and convert it to slope-intercept form (see Chapter 13). Thus, the equation of the perpendicular bisector of $\overline{DD'}$ is

$$y - 9 = 1(x - 8)$$
$$y = x + 1$$

Now do the same thing with $\overline{EE'}$. Its midpoint is $\left(\dfrac{15+15}{2}, \dfrac{-4+8}{2}\right)$, or (15, 2). $\overline{EE'}$ is vertical, so its perpendicular bisector is horizontal. The perpendicular bisector goes through (15, 2), so its equation is simply $y = 2$.

Finally, find the intersection of $y = 2$ and $y = x + 1$. That's (1, 2), the center of rotation. If you feel like it, locate (1, 2) on the figure, and then take a compass and place its point on (1, 2). You should be able to trace the circular arcs from *D* to *D'*, *E* to *E'*, and *F* to *F'*.

7 A clockwise rotation maps △*ABC* onto △*A'B'C'*. Find the center of rotation.

Tip: The math is a bit easier if you use $\overline{AA'}$ and $\overline{CC'}$.

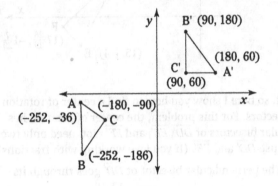

8 △*GHI* has been rotated 90° counterclockwise onto △*G'H'I'*. The origin is the center of rotation. Give the equations of *three* pairs of reflecting lines that would achieve the same result.

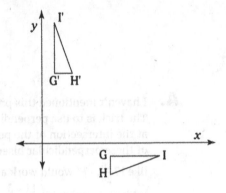

Working with Glide Reflections

A *glide reflection* is, as its name suggests, a glide (that's a translation) followed by a reflection (or vice versa). It's also referred to as a *walk*. See Figure 14-3.

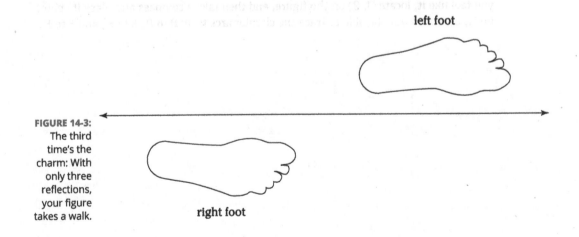

FIGURE 14-3: The third time's the charm: With only three reflections, your figure takes a walk.

left foot

right foot

How can you map the right foot onto the left? Well, you can't do it with a translation or a rotation, because translations and rotations don't change orientation, and you can see that these feet have opposite orientations. Reflections do reverse orientation, but there's no reflecting line that you can use to map the right foot onto the left. As you may suspect from the title of this section, the answer is that only a glide reflection can accomplish the mapping. You can map the right foot onto the left foot by reflecting the right foot over the line and then sliding it to the right (or by sliding it first, then reflecting it).

As you can see in the previous section on translations, you can achieve a translation or slide with two reflections. Thus, the glide part of a glide reflection can be done with two reflections. And that means that you can do a glide reflection — like the right foot to left foot mapping in Figure 14-3 — with only three reflections. And three reflections is the most you ever need to map a figure to another congruent figure. To sum up, any two congruent figures are always one reflection, two reflections (a translation or a rotation), or three reflections (a glide reflection) away from each other.

After you find the reflecting line for a glide reflection, the transformation is a cinch, because it's just a reflection (which you should already know how to do) followed by a translation in the direction of the reflecting line (which you also already know how to do). The following theorem tells you the key to finding the reflecting line.

Location of reflecting line in a glide reflection. In a glide reflection, the midpoints of all segments that connect pre-image points with their image points lie on the reflecting line.

Q. Find the reflecting line for the glide reflection.

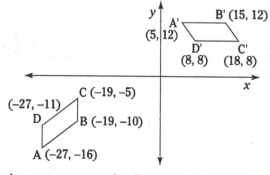

A. Pick any two point-image pairs, and make a segment out of each pair. Then, find the midpoints of these two segments. Next, find the slope of the line that goes through the two midpoints. (See Chapter 13 for info on slopes and midpoints.)

Using $\overline{AA'}$ and $\overline{BB'}$,

$$\text{Midpoint}_{\overline{AA'}} = \left(\frac{5-27}{2}, \frac{12-16}{2} \right) = (-11, -2)$$

$$\text{Midpoint}_{\overline{BB'}} = \left(\frac{15-19}{2}, \frac{12-10}{2} \right) = (-2, 1)$$

$$\text{Slope}_{\text{Reflecting Line}} = \frac{1-(-2)}{-2-(-11)} = \frac{1}{3}$$

Finally, just plug this slope and the point $(-2, 1)$ into the point-slope form for your equation:

$$y - 1 = \frac{1}{3}(x + 2)$$

That's all, folks.

---

*9 Use the transformation $T(x, y) \rightarrow (-x, y + 3)$.

a. Transform $\triangle LEG$ using $T(x, y)$. What are the new coordinates of L'', E'', and G''?

b. Find the equation of the reflecting line.

c. What are the coordinates of the image points (L', E', and G') obtained by reflecting L, E, and G over the line you found in part b, and what transformation, $T_{Reflect}(x, y)$, achieves this reflection?

d. After the reflection from part c is completed, what transformation, $T_{Glide}(x, y)$, completes the glide reflection?

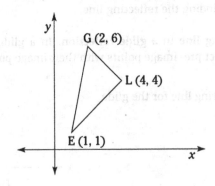

Solutions

(1) Here are the answers concerning orientation:

a. Opposite — you can't pair 'em up without a flip

b. Same

c. Same

d. These figures have neither the same nor opposite orientations because they're not congruent

e. Opposite

f. Opposite

(2) For $Q'R'S'T'$, here's what you get:

a.

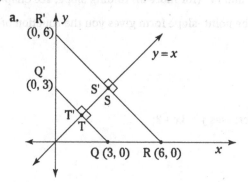

As you can see in the example problem, reflecting a figure over the line $y = x$ reverses the x- and y-coordinates of each point in the figure. Also note that S and S' are one and the same point. Ditto for T and T'. Any point that lies on the reflecting line stays put during a reflection.

b. $QQ'R'R$ is an isosceles trapezoid.

c. For the area of $QQ'R'R$, you could use the trapezoid area formula, but there's a much easier way. Call the origin point O. Now just subtract the area of right $\triangle OQ'Q$ from right $\triangle OR'R$:

$$\text{Area}_{QQ'R'R} = \text{area}_{OR'R} - \text{area}_{OQ'Q}$$
$$= \frac{1}{2}bh - \frac{1}{2}bh$$
$$= \frac{1}{2}(6)(6) - \frac{1}{2}(3)(3)$$
$$= 18 - 4.5$$
$$= 13.5 \text{ units}^2$$

d. $\triangle OQ'Q$ and $\triangle OR'R$ are $45° - 45° - 90°$ right triangles, so that makes figuring the perimeter of $QQ'R'R$ a snap:

$$\text{Perimeter}_{QQ'R'R} = QQ' + Q'R' + R'R + RQ$$
$$= 3\sqrt{2} + 3 + 6\sqrt{2} + 3$$
$$= 6 + 9\sqrt{2}$$

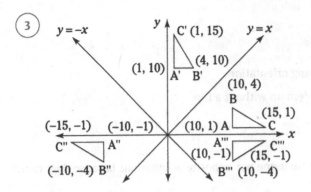

③

$y = -x$ C' (1, 15) $y = x$

(1, 10)   (4, 10)
A'   B'

(10, 4)
B

(15, 1)

(−15, −1)   (−10, −1)   (10, 1) A   C
C"   A"   A'''
(−10, −4) B"   (10, −1)   (15, −1)   C'''
B''' (10, −4)

④ Here's what happens when you reflect $\triangle TUV$:

a. $\overline{TT'}$ must be perpendicular to the reflecting line $y = 3x + 2$, which has a slope of 3. Thus, $\overline{TT'}$ has a slope of $-\frac{1}{3}$, as do $\overline{UU'}$ and $\overline{VV'}$ (for more on finding slope, see Chapter 13). Plugging $-\frac{1}{3}$ and $(8, -4)$ into the point-slope form gives you the equation of $\overline{TT'}$:

$$y - (-4) = -\frac{1}{3}(x - 8)$$

$$y = -\frac{1}{3}x - \frac{4}{3}$$

Next, find where this line, $\overline{TT'}$, crosses $y = 3x + 2$:

$$-\frac{1}{3}x - \frac{4}{3} = 3x + 2$$

$$-x - 4 = 9x + 6$$

$$-10x = 10$$

$$x = -1$$

And plugging this answer into $y = 3x + 2$ gives you $y = -1$. So, $\overline{TT'}$ crosses $y = 3x + 2$ at $(-1, -1)$. To get the coordinates of T', note that the reflecting line $y = 3x + 2$ must bisect $\overline{TT'}$, and thus $(-1, -1)$ must be the midpoint of $\overline{TT'}$. Going from T at $(8, -4)$ to $(-1, -1)$, you go left 9 and up 3. Do that again from $(-1, -1)$, and you get to T'. Left 9 from −1 brings you to −10, and up 3 from −1 brings you to 2. Thus, T' is at $(-10, 2)$.

In the interests of space, I'll skip the math for U' and V'. The procedure is identical to the one in the preceding paragraph. For the coordinates of U', you should get $(-5, 7)$, and for V', $(-6, 14)$.

b. You prove the triangles congruent with SSS, and to do that you just use the distance formula. Using the given coordinates of T, U, and V, you should get $5\sqrt{2}$, $5\sqrt{2}$, and $4\sqrt{10}$ for the lengths of the sides of $\triangle TUV$. And using the coordinates of T', U', and V' (which you calculated in part a), you should get the same three lengths for $\triangle T'U'V'$. That does it.

⑤ Answers vary. The translation is vertical, so the reflecting lines must be horizontal. And the lines have to be separated by half the length of $\overline{IT}$ (or $\overline{SR}$, $\overline{OA}$, or $\overline{CP}$), which is 5. Thus, any pair of horizontal lines separated by a distance of 2.5 will suffice. *Note:* The direction from l_1 to l_2 must be the same as the direction from the pre-image to the image.

Three possible answers are

- $l_1 : y = -1$ and $l_2 : y = 1.5$
- $l_1 : y = 2$ and $l_2 : y = 4.5$
- Or something crazy like $l_1 : y = -1002.5$ and $l_2 : y = -1000$

(6) Find the slope and midpoint of $\overline{CC'}$ ($\overline{AA'}$ and $\overline{BB'}$ would work just as well):

$$\text{Slope}_{\overline{CC'}} = \frac{0-(-2)}{9-0} = \frac{2}{9}$$

$$\text{Midpoint}_{\overline{CC'}} = \left(\frac{0+9}{2}, \ \frac{-2+0}{2}\right) = (4.5, -1)$$

You know l_1 and l_2 must be perpendicular to $\overline{CC'}$, so both lines have a slope of $-\frac{9}{2}$, or -4.5.
The first reflecting line, l_1, can go through C at $(0, -2)$:

$$y - (-2) = -4.5(x - 0)$$
$$y = -4.5x - 2$$

Then, l_2 would go through the midpoint of $\overline{CC'}$:

$$y - (-1) = -4.5(x - 4.5)$$
$$y = -4.5x + 19.25$$

You're done.

(7) You want to find the intersection of the perpendicular bisectors of $\overline{AA'}$ and $\overline{CC'}$. First, use the midpoint formula and the slope formula to compute the midpoint and slope of $\overline{AA'}$. You should get the following results:

$$\text{Midpoint}_{\overline{AA'}} = (-36, \ 12)$$
$$\text{Slope}_{\overline{AA'}} = \frac{2}{9}$$

The slope of the perpendicular bisector of $\overline{AA'}$ is the opposite reciprocal of the slope of $\overline{AA'}$, so its slope is $-\frac{9}{2}$, or -4.5. And thus, its equation is

$$y - 12 = -4.5[x - (-36)]$$
$$y = -4.5x - 150$$

Using the same method, you obtain the following for the equation of the perpendicular bisector of $\overline{CC'}$:

$$y = -1.8x - 96$$

The center of rotation lies at the intersection of these two perpendicular bisectors, so set the right sides of the equations equal to each other:

$$-4.5x - 150 = -1.8x - 96$$
$$-45x - 1,500 = -18x - 960$$
$$-27x = 540$$
$$x = -20$$

Plugging $x = -20$ into either equation gives you a y value of -60, so the center of rotation is at $(-20, -60)$.

(8) The rotation is 90° counterclockwise about the origin, so the reflecting lines must pass through the origin and form a 45° angle (half of 90°). Three possible answers are

- $l_1 : y = 0$ (the x-axis) and $l_2 : y = x$
- $l_1 : y = x$ and $l_2 : x = 0$ (the y-axis)
- $l_1 : x = 0$ and $l_2 : y = -x$

But any two lines work as long as they go through the origin and form a 45° angle. For example:

$$l_1 : y = \frac{3}{4}x \quad \text{and} \quad l_2 : y = 7x$$

(*9) Here's what happens with $\triangle LEG$:

a. $T(x, y) = (-x, y+3)$ sends points L, E, and G to the following image points:

$$L'' = T(4, 4) = (-4, 7)$$
$$E'' = T(1, 1) = (-1, 4)$$
$$G'' = T(2, 6) = (-2, 9)$$

b. You can use any two point-image pairs to find the reflecting line. How about $\overline{EE''}$ and $\overline{GG''}$?

$$\text{Midpoint}_{\overline{EE''}} = \left(\frac{1+(-1)}{2}, \frac{1+4}{2} \right) = (0, 2.5)$$

$$\text{Midpoint}_{\overline{GG''}} = \left(\frac{2+(-2)}{2}, \frac{6+9}{2} \right) = (0, 7.5)$$

Both midpoints are on the y-axis (if you realized that they would be before doing the math, you're a geometry natural), so the reflecting line must be the y-axis; its equation, of course, is $x = 0$.

c. Reflecting L, E, and G gives you

$$L' = (-4, 4)$$
$$E' = (-1, 1)$$
$$G' = (-2, 6)$$

The transformation that flips a figure over the y-axis is $T_{\text{Reflect}}(x, y) = (-x, y)$.

d. The transformation is just a slide straight up a distance of 3. That's achieved by $T_{\text{Glide}}(x, y) = (x, y+3)$.

IN THIS CHAPTER

» **Locating loci**

» **Locus hocus pocus**

» **Constructions: You'll need your compass and straightedge (and a hard hat)**

Chapter **15**

Laboring Over Loci and Constructions (No Proofs)

U p to this point in the book, you've been working on problems where you're given some shape (or shapes) — say, some lines, a triangle, a parallelogram, a circle — and you're asked to prove something about it, calculate something about it, or do something to it. But in this chapter, you have to come up with the geometric shape yourself. With locus problems, you're given certain conditions that the shape must satisfy, and you have to figure out what the shape is. And with construction problems, your task is to create the geometric object using only a compass and straightedge.

Tackling Locus Problems

Locus: A *locus* (plural *loci*) is a set of points (usually some sort of geometric object) consisting of all the points that satisfy certain conditions.

REMEMBER

Here's a simple example. What's the locus of all points 10 inches from a given point? The answer is a circle with a radius of 10 inches whose center is the given point.

When tackling a locus problem, it's a good idea to go through the following four-step process. This will help you avoid a couple of common mistakes: including too few points in your solution (see Step 2) and including too many points (see Step 3).

When working on locus problems, always follow this four-step method:

TIP

1. **Identify a pattern.**

   Sometimes you'll spot the key pattern right away. If so, you're done with Step 1. If not, find a single point that satisfies the given conditions; then find a second point, then a third, and so on until you recognize the pattern.

2. **Look outside the pattern for points to add.**

   Check for points outside the pattern you recognized in Step 1 that satisfy the given conditions. There might be isolated points or a significant geometric shape that you need to add to your locus solution.

3. **Look inside the pattern for points to exclude.**

   Check within the pattern you found in Step 1 to make sure that all the points within the pattern satisfy the conditions. If there are points you need to exclude, they're usually isolated points.

4. **Draw a diagram and write a description of the locus solution.**

Q. What's the locus of all points that are equidistant from two given points?

EXAMPLE **A.** Check out the following four-step solution.

1. **Identify a pattern.**

   Figure 15-1 shows the two given points, A and B, along with four new points that are each equidistant from the given points.

   See the pattern made by those four points? It's the vertical line that goes through the midpoint of the segment joining A and B. In other words, it's the perpendicular bisector of the segment.

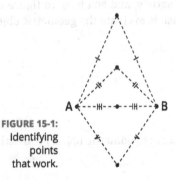

FIGURE 15-1:
Identifying
points
that work.

2. **Look outside the pattern for points to add.**

   Check for points outside the pattern you found in Step 1. You come up empty here. There's nothing to add.

3. **Look inside the pattern for points to exclude.**

   Ditto. No points need to be excluded.

4. **Draw a diagram and write a description of the locus solution.**

   Figure 15-2 shows the locus, and the caption gives its description.

FIGURE 15-2: The locus of points equidistant from two given points is the perpendicular bisector of the segment that joins the two points.

Q. What's the locus of all points that are equidistant from the following given intersecting lines?

EXAMPLE

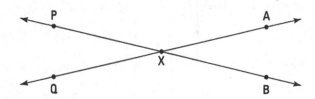

A. Do the four-step process.

1. **Identify a pattern.**

   Figure 15-3 shows four points that are equidistant from the two lines. Do you see the pattern? Right: Those four points lie on the ray shooting out to the right from X that bisects $\angle AXB$. The same thing works on the left side of X: That's the angle bisector of $\angle PXQ$. So, that gives you the line going through point X that bisects $\angle AXB$ and $\angle PXQ$. See Figure 15-4.

2. **Look outside the pattern for points to add.**

   You might have thought you were done, but this important Step 2 helps you see that you missed an entire second set of points. Do you see what you missed? It's the line through X that's perpendicular to the first line you identified. This second line bisects $\angle PXA$ and $\angle QXB$. See Figure 15-5.

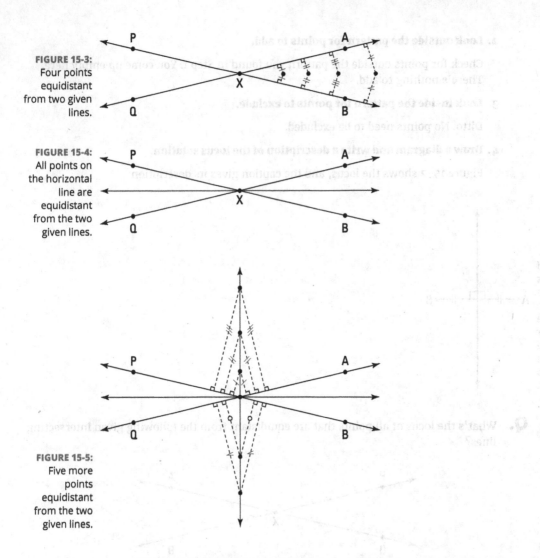

FIGURE 15-3: Four points equidistant from two given lines.

FIGURE 15-4: All points on the horizontal line are equidistant from the two given lines.

FIGURE 15-5: Five more points equidistant from the two given lines.

3. **Look inside the pattern for points to exclude.**

   Nothing to exclude.

4. **Draw the locus and describe it in words.**

   Figure 15-6 shows the locus solution.

   The locus is the perpendicular lines that intersect at X and that bisect the four angles made by the two given lines.

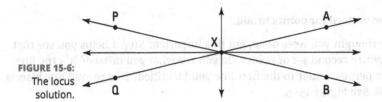

FIGURE 15-6: The locus solution.

1. What's the locus of the vertices of isosceles triangles having a given segment for a base?

2. What's the locus of all points 1 inch from a 1-inch-long segment, and what's the perimeter of this locus?

3. What's the locus of all points closer to the center of a given square than to any of the vertices of the square?

4. What's the locus of all the points in the x-y coordinate plane closer to the x-axis than the y-axis?

5 What's the locus of all points in the coordinate plane equidistant from the *x*-axis and the point (3, 1)?

Compass and Straightedge Constructions

As you may remember from a math class in middle school or junior high, the idea with construction problems is to construct geometric shapes using only a compass and a straightedge. I'm sure you know what a compass is, and a *straightedge* is — are you sitting down? — a straight edge. It's basically a ruler (and you can use a ruler for these problems), but a straightedge has no length marks on it. So, if you use a ruler, you're not allowed to use any of its markings to measure anything.

In the example problems and in the solutions to the practice problems, I use the following notation. To indicate where to draw an arc with your compass, I first name the point where you put the point of the compass (this is the center of the circular arc), and then I write how wide you should open the compass (this is the radius of the arc). The radius could be listed as the length of a segment or with a single letter. For example, arc (*Q*, *QP*) is the arc with center at point *Q* and a radius that's the length of segment *QP*, and arc (*X*, *r*) is the arc with center at point *X* with a radius of *r*.

EXAMPLE

Q. Given: ∠*A*.

Construct: ∠*B* that's congruent to ∠*A*.

A. Refer to Figure 15-7 as you go through the following steps.

1. Draw a working line, *l*, with point *B* on it.

2. Open your compass to any radius *r*, and construct arc (*A*, *r*) intersecting the two sides of ∠*A* at points *S* and *T*.

3. Construct arc (*B*, *r*) intersecting line *l* at some point *V*.

4. Construct arc (*S*, *ST*).

5. Construct arc (*V*, *ST*) intersecting arc (*B*, *r*) at point *W*.

6. Draw $\overline{BW}$. That does it. (Note: I didn't show Step 6 in the figure because it would make the figure a bit confusing.)

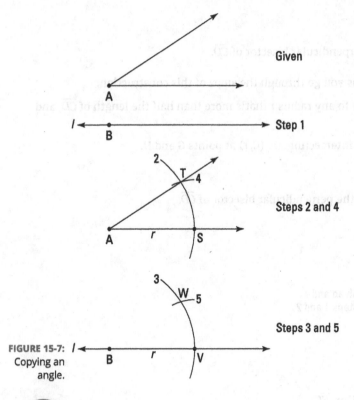

Given

Step 1

Steps 2 and 4

Steps 3 and 5

FIGURE 15-7:
Copying an
angle.

Q. Given: $\angle K$.

Construct: $\overline{KZ}$, the bisector of $\angle K$.

EXAMPLE

A. Refer to Figure 15-8 as you go through this construction.

1. Open your compass to any radius r, and construct arc (K, r) intersecting the two sides of $\angle K$ at A and B.

2. Use any radius s to construct arc (A, s) and arc (B, s) that intersect each other at point Z.

Note that you must choose a radius s that's long enough for the two arcs to intersect.

3. Draw $\overline{KZ}$. That's a wrap.

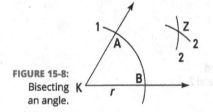

Given and
Steps 1 and 2

FIGURE 15-8:
Bisecting
an angle.

Q. Given: $\overline{CD}$.

EXAMPLE

Construct: $\overleftrightarrow{GH}$, the perpendicular bisector of $\overline{CD}$.

A. Refer to Figure 15-9 as you go through the steps of this construction.

1. Open your compass to any radius r that's more than half the length of $\overline{CD}$, and construct arc (C, r).

2. Construct arc (D, r) intersecting arc (C, r) at points G and H.

3. Draw $\overleftrightarrow{GH}$.

You're done. $\overleftrightarrow{GH}$ is the perpendicular bisector of $\overline{CD}$.

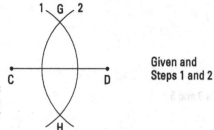

Given and
Steps 1 and 2

FIGURE 15-9:
Constructing a
perpendicular
bisector.

Q. Given: $\overleftrightarrow{EF}$ and point W on $\overleftrightarrow{EF}$.

EXAMPLE

Construct: $\overleftrightarrow{WZ}$ such that $\overleftrightarrow{WZ} \perp \overleftrightarrow{EF}$.

A. Figure 15-10 shows the steps of this construction.

1. Using any radius r, construct arc (W, r) that intersects $\overleftrightarrow{EF}$ at X and Y.

2. Using any radius s that's greater than r, construct arc (X, s) and arc (Y, s) intersecting each other at point Z.

3. Draw $\overleftrightarrow{WZ}$.

That's it. $\overleftrightarrow{WZ}$ is perpendicular to $\overleftrightarrow{EF}$ at point W.

FIGURE 15-10:
Constructing a
perpendicular
line through a
point on a line.

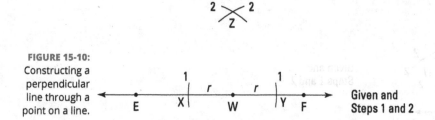

Given and
Steps 1 and 2

EXAMPLE

Q. Given: $\overline{AZ}$ and point J not on $\overline{AZ}$.

Construct: $\overline{JM}$ such that $\overline{JM} \perp \overline{AZ}$.

A. Refer to Figure 15-11.

  1. Open your compass to a radius r that's greater than the distance from J to $\overline{AZ}$, and construct arc (J, r) intersecting $\overline{AZ}$ at K and L.

  2. Leaving your compass open to radius r, construct arc (K, r) and arc (L, r) — on the side of $\overline{AZ}$ that's opposite point J — intersecting each other at point M.

  3. Draw $\overline{JM}$. You're done.

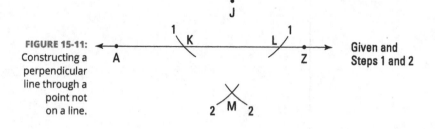

FIGURE 15-11: Constructing a perpendicular line through a point not on a line.

Given and Steps 1 and 2

---

6 Construct a triangle whose sides are in the ratio of $2.1 : 2.8 : 3.5$. (*Hint:* What type of triangle is that?)

For problems 7 and 8, use $\triangle ABC$ shown in Figure 15-12.

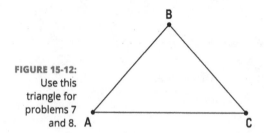

FIGURE 15-12: Use this triangle for problems 7 and 8.

7 Construct △*XYZ* that's congruent to △*ABC*.

8 Construct the incenter of △*ABC*.

Solutions

(1) Here's the four-step solution.

1. Identify a pattern.

Look back at Figure 15-1 that goes with the first example problem. You can see three isosceles triangles with base $\overline{AB}$. So, you might think this locus problem has the very same solution as example problem 1, namely, the perpendicular bisector of $\overline{AB}$. However. . .

2. Look outside the pattern for points to add.

This is a bit tricky. This locus problem asks for all points that are vertices of the isosceles triangles with base $\overline{AB}$. Well, all triangles have three vertices. Points A and B are two of the vertices of the three triangles you see in Figure 15-1; and they are, of course, vertices of all isosceles triangles with base $\overline{AB}$. Thus, you must add points A and B to the perpendicular bisector of $\overline{AB}$ identified in Step 1. And. . .

3. Look inside the pattern for points to exclude.

Warning: Don't neglect Steps 2 and 3! In this particular locus problem, both of these steps are critical.

There's a single point that must be excluded from the solution. Did you find it? It's the midpoint of $\overline{AB}$. Except for this midpoint, all points along the perpendicular bisector of $\overline{AB}$ form a triangle with points A and B. But the midpoint of $\overline{AB}$ is on the same line as points A and B, and three collinear points cannot be the vertices of a triangle.

4. Draw the locus and describe it in words.

The locus is the perpendicular bisector of $\overline{AB}$ plus points A and B minus the midpoint of $\overline{AB}$.

The diagram of the locus is the same as Figure 15-1 with the addition of a hollow dot where the perpendicular bisector intersects $\overline{AB}$; and it would have to be made clear that points A and B are included in the locus solution.

(2a) I think you know what to do: Four steps, naturally.

1. Identify a pattern.

Look at Figure 15-13.

You can see that points one inch above and one inch below $\overline{JK}$ satisfy the locus condition. Those points are one inch straight down to $\overline{JK}$ or one inch straight up to $\overline{JK}$. But for a point to the left of point J, its distance to $\overline{JK}$ is the distance to endpoint J. Such points form a semicircle with center at J and a radius of one inch. See Figure 15-14.

The same thing applies, of course, to the right of point K.

2, 3. Look inside and outside the pattern.

Steps 2 and 3 yield no changes to the pattern found in Step 1.

4. Draw the locus and describe it in words.

The locus is an oval path consisting of two segments and two semicircles. See Figure 15-15.

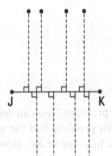

FIGURE 15-13:
Points an inch
above and an
inch below $\overline{JK}$.

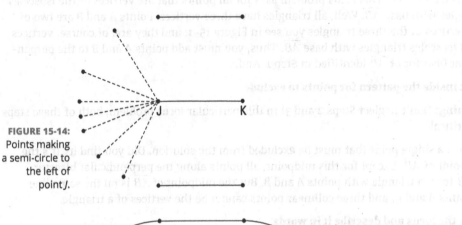

FIGURE 15-14:
Points making
a semi-circle to
the left of
point J.

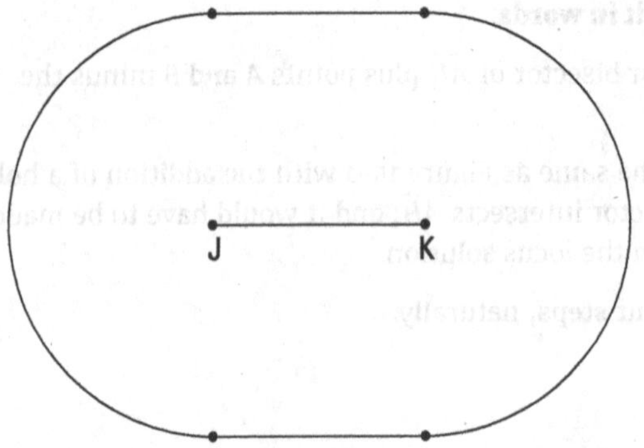

FIGURE 15-15:
The locus
solution.

(2b) The perimeter is made up of two one-inch segments and two semicircles with a one-inch radius. That gives you a perimeter of $2 + 2\pi$.

(3) Four steps as usual:

1. Identify a pattern.

Consider the perpendicular bisector of $\overline{AX}$. See Figure 15-16.

Points on this perpendicular bisector are equidistant from A and X. Thus, to be closer to X than to A, a point must be on the lower-right side of the perpendicular bisector. The same argument applies to the perpendicular bisectors of $\overline{BX}$, $\overline{CX}$, and $\overline{DX}$. See Figure 15-17.

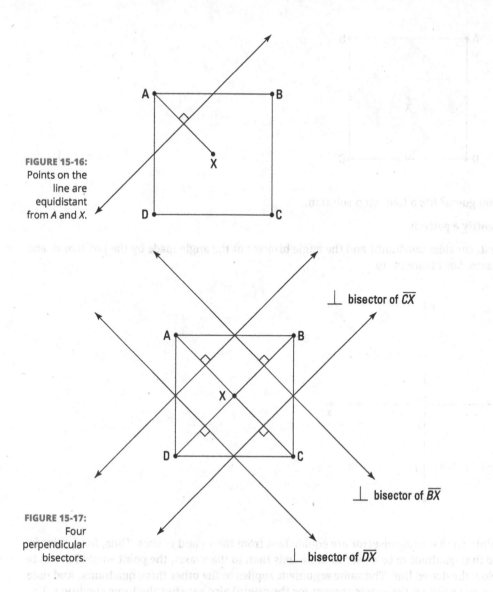

FIGURE 15-16:
Points on the
line are
equidistant
from *A* and *X*.

⊥ bisector of $\overline{CX}$

⊥ bisector of $\overline{BX}$

FIGURE 15-17:
Four
perpendicular
bisectors.

⊥ bisector of $\overline{DX}$

2, 3. **Look inside and outside the pattern.**

Steps 2 and 3 yield no changes to the pattern found in Step 1.

4. **Draw the locus and describe it in words.**

The locus is the interior of a square with vertices at the midpoints of the sides of the original square. Note that the sides of this square are dotted line segments, which indicates that the sides of this square are not part of the locus solution. See Figure 15-18.

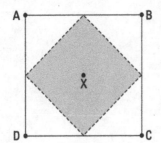

FIGURE 15-18:
The final result.

④ Can you guess? It's a four-step solution.

1. Identify a pattern.

First, consider quadrant I and the angle bisector of the angle made by the positive x- and y-axes. See Figure 15-19.

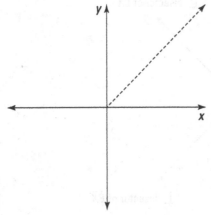

FIGURE 15-19:
A 45° ray in quadrant I.

Points on that angle bisector are equidistant from the x- and y-axes. Thus, for a point in the first quadrant to be closer to the x-axis than to the y-axis, the point would have to be below the dotted line. The same argument applies to the other three quadrants. And note that any point on the x-axis (except for the origin) also satisfies the locus condition. The result is the shaded region shown in Figure 15-20.

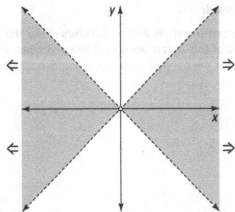

FIGURE 15-20:
A bow tie of sorts.

2, 3. Look inside and outside the pattern.

Steps 2 and 3 yield no changes to the pattern found in Step 1.

4. Draw the locus and describe it in words.

The bow-tie-shaped shaded area shown in Figure 15-18 goes forever to the right and to the left. And note that the borders are dotted lines because points on the lines are equidistant from the axes and, thus, are not part of the locus solution. (Also note that the equations of the dotted border lines are $y = x$ and $y = -x$.)

⑤ The four-step process isn't necessary or particularly helpful for this problem.

Identifying a pattern doesn't work as well here as with the other problems because the answer isn't some simple shape like a line or a circle or a square. But you might recognize the shape after putting down five points that satisfy the locus condition at $\left(3, \frac{1}{2}\right)$, $(2, 1)$, $(4, 1)$, $(0, 5)$, and $(6, 5)$. (For these last two points, do you see how you can use a $3 - 4 - 5$ triangle to locate them?) See Figure 15-21.

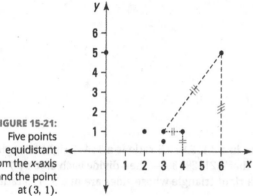

FIGURE 15-21:
Five points equidistant from the *x*-axis and the point at $(3, 1)$.

Do you recognize this shape? That's it — it's a parabola. But to determine the precise locus solution, you need to solve the problem algebraically.

Consider a general point in the coordinate plane, (x, y). (Note that to satisfy the locus condition, this point must be above the *x*-axis.) Its distance from the *x*-axis is simply *y*, and its distance from $(3, 1)$ is given by the distance formula:

$$d = \sqrt{(x-3)^2 + (y-1)^2}$$

Set that equal to y and solve:

$$y = \sqrt{(x-3)^2 + (y-1)^2}$$
$$y^2 = (x-3)^2 + (y-1)^2$$
$$y^2 = x^2 - 6x + 9 + y^2 - 2y + 1$$
$$2y = x^2 - 6x + 10$$
$$y = \tfrac{1}{2}x^2 - 3x + 5$$

Points on that parabola satisfy the locus condition. Figure 15-22 shows the final solution.

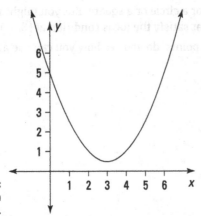

FIGURE 15-22:
The parabola
$y = \tfrac{1}{2}x^2 - 3x + 5$.

(6) Did you realize that a triangle whose sides are in a $2.1 : 2.8 : 3.5$ ratio is a $3 : 4 : 5$ right triangle? To see that, first multiply each side by 10 — that's $21 : 28 : 35$ — then divide each side by 7 — $3 : 4 : 5$. So, all you need to do is to construct a right triangle whose sides are in a $3 : 4 : 5$ ratio. Piece o' cake.

First, construct a right angle at point A using the technique explained in the fourth example problem. That will give you the perpendicular lines shown in Figure 15-23.

Next, simply use your compass to mark off four arcs along the horizontal line and three arcs along the vertical line. The fourth arc on the horizontal line gives you vertex B, and the third arc on the vertical line gives you vertex C. Connect B and C and you're done. $\triangle ABC$ is a $3 - 4 - 5$ right triangle whose sides are in a ratio of $2.1 : 2.8 : 3.5$.

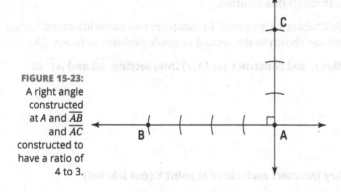

(7) Refer to Figure 15-24 as you go through the following steps.

1. **Draw a working line, *l*, with a point *X* on it.**

2. **Open your compass to the length of $\overline{AC}$, and then construct $\overline{XZ}$ on line *l* that's the same length as $\overline{AC}$.**

3. **Construct the following:**

   a. arc (*A*, *AB*)

   b. arc (*X*, *AB*)

4. **Construct the following:**

   a. arc (*C*, *CB*)

   b. arc (*Z*, *CB*) intersecting arc (*X*, *AB*) at point *Y*

5. **Draw $\overline{XY}$ and $\overline{ZY}$ and you're done.**

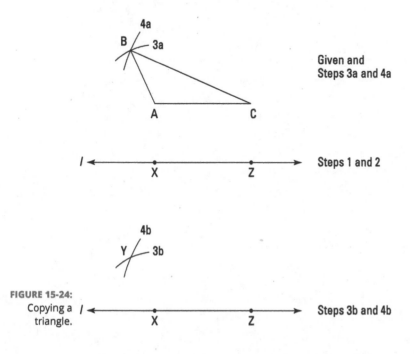

⑧ Refer to Figure 15-25 as you work through this solution.

To locate the incenter of △ABC (see Chapter 4), you need to construct two angle bisectors. Let's bisect angles A and C. Use the technique shown in the second example problem to bisect ∠A.

1. **Open your compass to any radius r, and construct arc (A, r) intersecting $\overline{AB}$ and $\overline{AC}$ at points P and Q.**

2. **Construct the following:**

   a. arc (P, r) and

   b. arc (Q, r)

3. **Construct these arcs so that they intersect each other at point X (not labeled).**

4. **Draw $\overrightarrow{AX}$. That's the bisector of ∠A.**

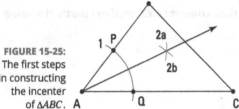

FIGURE 15-25:
The first steps in constructing the incenter of △ABC.

Repeat this process to construct the angle bisector of ∠C. The two angle bisectors intersect at the incenter of △ABC.

Chapter **16**

Ten Things You Better Know (for Geometry), or Your Name Is Mudd

actually don't have any problem with people named Mudd (for all you Mudds out there who are reading this book), but if you don't know these things, you really should go back and look through this book again! You need all the formulas and theorems in this chapter if you really want to be an expert in "the study of shapes."

The Pythagorean Theorem (the Queen of All Geometry Theorems)

The sum of the squares of the legs of a right triangle is equal to the square of the hypotenuse, or

$a^2 + b^2 = c^2$ (See Chapter 4.)

Special Right Triangles

The first four triangles in this section are so-called Pythagorean triple triangles. They're special because the lengths of all three sides are integers, which doesn't happen often with the Pythagorean Theorem (usually you get a square root of something for at least one of the sides):

» The $3-4-5$ triangle

» The $5-12-13$ triangle

» The $7-24-25$ triangle

» The $8-15-17$ triangle

The next two triangles are special because they're related to two of the most basic shapes in geometry: The first is half of a square, and the second is half of an equilateral triangle. They come up all the time in problems, so make sure you know them! (See Chapter 4 for details.)

» The $45°-45°-90°$ triangle, whose sides are in the ratio of $x : x : x\sqrt{2}$

» The $30°-60°-90°$ triangle, whose sides are in the ratio of $x : x\sqrt{3} : 2x$

Area Formulas

The following formulas give you the area of triangles and special quadrilaterals (see Chapter 7):

» $\text{Area}_{\text{Triangle}} = \frac{1}{2} \text{ base} \cdot \text{height}$

» $\text{Area}_{\text{Parallelogram}} = \text{base} \cdot \text{height}$

(This formula also works for rectangles and squares because they're parallelograms.)

» $\text{Area}_{\text{Kite}} = \frac{1}{2} \text{diagonal}_1 \cdot \text{diagonal}_2$

(This formula also works for rhombuses and squares because they're kites.)

» $\text{Area}_{\text{Trapezoid}} = \frac{\text{base}_1 + \text{base}_2}{2} \cdot \text{height}$

Sum of Angles

The sum of the *interior* angles of a polygon with n sides is $(n-2)180°$. The sum of the *exterior* angles of any polygon is $360°$. (See Chapter 7 for more information.)

Circle Formulas

Try these equations (which you can find in Chapter 10) when you work with circumference and area:

» Circumference $= 2\pi r = \pi d$

» $\text{Area}_{\text{Circle}} = \pi r^2$

Angle-Arc Theorems

In some circle problems, you can have an angle whose vertex is *on* the circle or whose vertex is *outside* the circle or whose vertex is *inside* the circle. The following formulas give you the connection between the size of the angle and the arc it intercepts (see Chapter 10). Figure 16-1 gives examples of the types of angles these formulas apply to:

» Angle *on* a circle $= \frac{1}{2} arc_1$

» Angle *outside* a circle $= \frac{1}{2}\left(arc_2 - arc_3\right)$

» Angle *inside* a circle $= \frac{1}{2}\left(arc_4 + arc_5\right)$

Note: You get an angle inside a circle when two chords cross each other, forming an X; for this formula, you use the arcs intercepted by the angle you want and its vertical angle.

FIGURE 16-1:
Angles (a) on, (b) outside, (c) inside a circle.

 a) 1

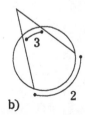 b) 2

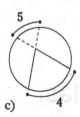 c) 4

Power Theorems

Memorize the following theorems and become a geometry powerhouse (see Chapter 10):

» **Chord-chord:** part · part = part · part

» **Secant-secant:** whole · outside = whole · outside

» **Secant-tangent:** whole · outside = tangent^2

All three of these theorems follow the same simple rule:

$$(\text{vertex to circle}) \cdot (\text{vertex to circle}) = (\text{vertex to circle}) \cdot (\text{vertex to circle})$$

Coordinate Geometry Formulas

Given two points in the coordinate plane, (x_1, y_1) and (x_2, y_2), you can compute the slope between the two points, the halfway point between the points, and the distance from one point to the other with the following formulas (see Chapter 13):

» Slope $= \dfrac{y_2 - y_1}{x_2 - x_1}$

» Midpoint $= \left(\dfrac{x_1 + x_2}{2}, \dfrac{y_1 + y_2}{2} \right)$

» Distance $= \sqrt{(x_2 - x_1)^2 + (y_2 - y_1)^2}$

Volume Formulas

Here's how to find the volume of spheres, flat-top solids like cylinders and prisms, and pointy-top solids like pyramids and cones (see Chapter 12):

» $\text{Vol}_{\text{Sphere}} = \dfrac{4}{3} \pi r^3$

» $\text{Vol}_{\text{Flat-Top solids}} = \text{area}_{\text{base}} \cdot \text{height}$

» $\text{Vol}_{\text{Pointy-Top solids}} = \dfrac{1}{3} \text{area}_{\text{base}} \cdot \text{height}$

Surface Area Formulas

And here's how to find the surface area of spheres, flat-top solids, and pointy-top solids (see Chapter 12):

» $\text{SA}_{\text{Sphere}} = 4 \pi r^2$

» $\text{SA}_{\text{Flat-Top solids}} = 2 \cdot \text{area}_{\text{base}} + \text{lateral area}_{\text{rectangle(s)}}$

» $\text{SA}_{\text{Pointy-Top solids}} = \text{area}_{\text{base}} + \text{lateral area}_{\text{triangle(s)}}$

Index

Corresponding Parts of Congruent
Triangles are Congruent (CPCTC),
97–101, 115–117

Corresponding Sides of Similar Triangles
are Proportional (CSSTP), 183–186,
198–200

cylinders, 253–255, 261–263

D

Descartes, René, 271

diagonals, in polygons, 166

diameter, 208

distance, and chord size, 208

distance formula, 272–274, 279–282, 322

dividing angles and segments, 42–46, 52–53

Dunce Cap Theorem, 215, 222

E

equations of lines and circles, 276–278,
283–284

equiangular polygons, 166

equiangular triangles, 58

equidistance theorems, 108–112, 120

equilateral triangles, 58–62, 80–81, 163–164

Example icon, explained, 3

exterior angles in polygons, 165–167,
172–173, 320

external tangent, 216

F

families, triangle, 75

flat-top figures, 253–255, 261–263, 322

flexible approach to proofs, 116

flow-of-logic structure in proofs, 13–14, 33

45°-45°-90° triangles, 78–79, 88–90, 320

G

general principles, in proofs, 13

geometry
defined, 7
essentials, 319–322
making right assumptions in, 8–10
overview, 1–4
proofs overview, 12–14
solutions to practice problems, 15–16
theorems and postulates in, 11–12

givens in proofs, importance of using, 38,
39, 98

glide reflections, 294–296, 300

H

height
slant, 256, 257, 263
of triangles, 61–64, 81–82

Hero's formula, 62, 82

horizontal lines, 272, 277

hypotenuse
Altitude-on-Hypotenuse Theorem,
186–189, 200–202
defined, 58
in 45°-45°-90° triangles, 78, 79

Hypotenuse-Leg (HL) method, 105–108,
119–120

I

icons, explained, 3

if angles, then sides, 102–104, 117–118

if sides, then angles, 102–104, 117–118, 196

if-then logic, in proofs, 12–14, 34

incenter, of triangles, 66–70, 83–85

interior angles, in polygons, 165–167,
172–173, 320

internal tangent, 216

intersecting lines and planes, 245–248,
251–252

intersection problems, 18–20, 28

isometries
glide reflections, 294–296, 300
overview, 285
reflections, 286–289, 297–298
rotations, 292–294, 299–300
translations, 289–291, 298–299

isosceles trapezoids, 129, 141–143, 156

isosceles triangles
altitude of, 62
overview, 58–61, 80–81
rules for, using to prove congruent
triangles, 102–104, 117–118

K

kites
area formula for, 160, 171, 320
overview, 128
properties of, 132–137, 152–154
proving, 143–147, 157–158

L

lateral area, 254, 256

legs, in triangles, 58

Like Divisions Theorem, 42, 44

Like Multiples Theorem, 42, 43, 118

line segments. *See* segments

lines
defined, 17–18
equations of, 276–278, 283–284
horizontal, 272, 277
intersecting, 245–248, 251–252
parallel
and planes, 245–248, 251–252
slope of, 272
theorems involving proportions, 190, 202
transversals and, 124–128, 149–150

perpendicular
multiple, 245–248, 251–252
to planes, 241–244, 249–250
slope of, 272
slope formulas, 272–274, 279–282
vertical, 272, 277

lines-cut-by-a-transversal theorems, 125

locus problems, 301–306, 311–316

M

major arc, 211

medians
of trapezoids, 160
of triangles, 65–66, 83

midpoint, defined, 21–22, 28–29

midpoint formula, 272–274, 279–282, 322

minor arc, 211

multiplying angles and segments, 42–46,
52–53

O

obtuse angles, 18

obtuse triangles
altitude of, 62
formed by parallel lines and transversals,
124
orthocenter and circumcenter in, 67
overview, 58–61, 80–81

online practice material, accessing, 4

orientation, reflections and, 286, 288, 297

orthocenter, of triangles, 66–70, 83–85

P

parallel, defined, 23

parallel lines
and planes, 245–248, 251–252
slope of, 272
theorems involving proportions,
190, 202
transversals and, 124–128, 149–150

parallelograms
area formula for, 160, 168, 171, 320
overview, 128
properties of, 132–137, 152–154
proving, 143–147, 157–158

patterns, in locus problems, 302–304,
311–315

perimeter
of circles, 223–226, 233–236
of similar polygons, 176–177

perpendicular bisectors, 109–111, 287–288,
293, 308

perpendicular lines
multiple, 245–248, 251–252
to planes, 241–244, 249–250
slope of, 272

About the Author

A graduate of Brown University and the University of Wisconsin Law School, **Mark Ryan** has been teaching math since 1989. He runs The Math Center in Winnetka, Illinois (www.themathcenter.com), a one-man math teaching and tutoring business; he helps students with all junior high and high school math courses, including calculus and statistics, as well as ACT, PSAT, and SAT preparation. In high school, Ryan twice scored a perfect 800 on the math portion of the SAT, and he not only knows mathematics, but he also has a gift for explaining it in plain English. He practiced law for four years before deciding he should do something he enjoys and use his natural talent for mathematics.

Geometry Workbook For Dummies, 2nd Edition, is Mark Ryan's thirteenth book. His first book, *Everyday Math for Everyday Life* (Grand Central Publishing), was published in 2002. For Wiley, *Calculus For Dummies*, 1st Edition, was published in 2003; *Calculus Workbook For Dummies*, 1st Edition, in 2005; *Geometry Workbook For Dummies*, 1st Edition, in 2007; *Geometry For Dummies*, 2nd Edition, in 2008; *Calculus Essentials For Dummies* in 2010; *Geometry Essentials For Dummies* in 2011; *Calculus For Dummies*, 2nd Edition, in 2014; *Calculus Workbook For Dummies*, 2nd Edition, in 2015; *Geometry For Dummies*, 3rd Edition, in 2016; *Calculus Workbook For Dummies*, 3rd Edition, in 2018; and *Calculus All-In-One For Dummies* in 2023. Ryan's math books have sold over 875,000 copies.

Ryan lives in Evanston, Illinois. For fun, he hikes, skies, plays platform tennis, travels, plays on a pub trivia team, and roots for the Chicago Blackhawks and the Cubs.

Dedication

To my current and former math students. Through teaching them, they taught me.

Author's Acknowledgments

This book is a testament to the high standards of everyone at Wiley. I've worked for many years with Executive Editor Lindsay Berg. She's intelligent, professional, down-to-earth, and has a great sense of humor. She always has an empathetic and quick understanding of my concerns as an author. And she has a special way of dealing with my oh-so-minor personality foibles with patience, skill, and finesse. It's always a pleasure to work with her. Kristie Pyles, Senior Managing Editor, managed the production schedule and page proofs process. Many thanks to Kristie for being the ultimate manager from the beginning to the very end. Authors might want to polish their books forever, but Kristie ensures that they get done. Marylouise Wiack, Copy Editor, has been invaluable in proofreading and editing the entire book, including the extremely complicated mathematics. She has a great eye for detail and precision. Amy Nicklin, Technical Editor, did an excellent job checking the many hundreds of equations in the book. She's a math expert with a great eye for spotting errors. The layout and graphics team did a fantastic job with the book's complex equations and mathematical figures. Finally, I've done five books with Development Editor Tim Gallan. He's a very talented and experienced editor who has a deft touch with the many aspects of taking a book from start to finish. He understands the forest, the trees, when to edit, and when not to edit. And his laid-back style fits well with my . . . uh . . . not-so-laid-back style. I'm so grateful for all this help from such talented professionals.

Publisher's Acknowledgments

Executive Editor: Lindsay Berg

Senior Managing Editor: Kristie Pyles

Development Editor: Tim Gallan

Copy Editor: Marylouise Wiack

Technical Editor: Amy Nicklin

Production Editor: Saikarthick Kumarasamy

Cover Image: © Andrei Akushevich/Getty Images